A Guide to the Electronic Document Body of Knowledge

(EDBOK™ Guide)

McGrewGroup LLC / MC2 Books
Aurora, Colorado

Library of Congress Control Number:

ISBN: 978-1-893347-05-2

Jacket & Book Design by MC2Designs

First Printing, August 2014

Contents

Acknowledgments

Xplor International and the Xplor EDP Commission would like to express their appreciation to those individuals who have contributed their time, knowledge, expertise, and tenacity to creating the first edition of ***A Guide to the Electronic Document Body of Knowledge (EDBOK™ Guide)***. The EDBOK forms the basis for the Electronic Document Professional® certification program which awarded the first EDPs in 1990 and continues today. The EDP® program is administered through Xplor International.

The concept for a published EDBOK was initiated by the EDP Commission and we would like to recognize the project team for the first edition and their tireless efforts in steering the project through to completion:

- Matt Riley edp, Project Chair
- Neil Merchant m-edp, Project Manager
- Pat McGrew m-edp, Editor
- Scott Baker
- Roberta McKee-Jackson edp
- Chad Henk eda

We would like to acknowledge the technology Contributors to the EDBOK including:

- Paul Abdool edp
- William Broddy m-edp
- Franklin Campbell edp
- Tim Ciceran
- Brett Dashwood edp
- Christine Durfee
- Carol Fiore
- Franklin Friedmann edp
- Neal Gottsacker
- Chris Halicki edp
- Cheryl Kay
- Kevin Lantaff edp
- Robert Linsky
- Wendy MacMillan m-edp
- William McCalpin m-edp
- Linda McDaniel edp
- Roberta McKee-Jackson edp
- Neil Merchant m-edp
- Denise Miano edp
- Tim Nelms
- Stephen Poe edp
- Rebecca Rodgers m-edp
- Kevin Tondreau edp

This first edition of ***A Guide to the Electronic Document Body of Knowledge (EDBOK™ Guide)*** is a continuning work in progress and we encourage readers to help us expand and improve the content. If you find omissions, inaccuracies or ambiguities, please advise the EDBOK project team at edp@xplor.org. Your observations will help improve future editions of the ***EDBOK Guide***.

Roberta McKee-Jackson edp

Chair, EDP Commission

Overview

Contributed by Neil Merchant m-edp

Why a Body of Knowledge?

As in most skill and knowledge domains, there are two major categories of work: developing and operating. Be it sales, product development, print-for-profit or in-house billing work, we have to first design and implement a solution and a process, and then we operate and maintain them.

In both categories, there is a chronological sequence of activities. The EDBOK is structured to reflect this chronology. Firstly, it describes the document systems development lifecycle, then goes on to describe the document production workflow. What follows here is a précis of each of the topics that comprise the main body of the publication, to help set them in context and to summarise the flows.

Xplor's roots lie in the sharing of technical experience and expertise. The EDBOK reflects that technical heritage, and one of our challenges is to broaden out our knowledge base to encompass other aspects of document industry expertise. It's our belief, though, that while the activities may have different names and characteristics they're pretty much universal. So, feel free to impose your own skillset, work and experience on these topics if necessary, to make them meaningful for you and to help you navigate the flows.

Document Systems Development Lifecycle

This section describes the process of developing a new or changed transaction document-related solution. There are many development methodologies, varying by and within industries. The names of the phases described in the topics in this section of the EDBOK are those that we in Xplor have adopted for our body of knowledge definitions. You may find that the methodology you follow uses different names and locates some activities in different phases. Regardless, you should be able to recognize their purposes and interdependencies, and relate them to your own experience.

Requirements Gathering

During this first phase of the project, the goal is, at a high level, to define the project's purpose, core needs, scope, and stakeholders. The primary deliverable is a clear project definition that serves as a baseline explanation and agreement for all project participants.

Business Analysis

During this phase, detailed investigations take place to document the changes in business processes (manual and automated), and develop the project cost-benefit analysis based on current costs, forecast cost-savings, new costs, business benefits, costs of conversion, and potential costs of inertia.

Technical Analysis

During this Phase (which is usually concurrent to Business Analysis), a technical investigation into detailed business requirements is conducted. This identifies the current state of technology or process, the desired state of technology or process, the gaps to be closed, and the anticipated project path to the desired state.

At the end of business and technical analysis, the detailed requirements (what functionality is required) should be clearly understood and documented.

Stakeholder Agreement

Stakeholder engagement and agreement is key to the success of any project; indeed, without clear stakeholders there should BE no project. However, at this point in the lifecycle, when the problem is clearly defined and the cost-benefit analysis is more confident, stakeholders can commit (or not) to supporting its progress through to implementation.

Architecture

This and the subsequent step, Design, are often combined into one. Architecture defines the high-level structure of the solution, and what software, hardware, CRM or other platforms (typically an organization's strategic tool set for this kind of deliverable) should be used. Either here or in Design, document integrity controls will be specified, spanning every step of the production process.

Information Design

In this phase, included in the project if a new or redesigned document is involved, the document is designed to convey the necessary information clearly and effectively. Information design is the defining, planning, and shaping of the contents of a message and the environments (paper, web, mobile, special needs) in which it is presented, with the intention of satisfying the information needs of the intended recipients. Unlike most other phases in the development lifecycle, this is not technical work, and is best addressed by a specialist in the field.

(Project) Design

During this phase, program, data and job-flow designs, document factory layouts and workflows, or other specifications are produced to a level of detail that will enable the next step, Development (Build) to proceed. During Design, final document layouts may be devised, the data dictionary data-elements specified, and all document objects identified.

Development

During the Development (or Build) phase, programming work, hardware installation and equivalent activities take place.

Critical Communications Recovery (Disaster Recovery)

During this phase, plans, procedures and services are put in place to ensure that, in the event of a major disruption to business operations (e.g. fire, flood, sustained power or telecommunications outage), the production of documents and the cashflow, legal compliance, customer service and brand image they represent is preserved.

Test and QA

During this phase the built solution from the previous phase is subjected to testing to ensure that it works as designed, performs to expectations, and meets client requirements. This can include, for example: test case development and execution; paper and environmental testing; data to post office testing, and parallel processing.

Production (Launch)

The tested and proven solution is now moved into production/operational use. Many activities are involved: initial staff training, often across multiple areas; preparation of operational instructions; documenting ongoing management of process; staffing and personnel management; logistics planning; paper and environmental monitoring, and service level management. At the end of this phase, the solution is implemented and delivering its benefits to the organization.

Maintenance

A newly implemented solution often comes with a warranty period, during which any problems will be resolved free of charge. Maintenance involves: error correction (not meeting all the documented and agreed requirements, or failing in some manner); ongoing training; online or call centre technical support; continuous process improvement, and ultimate decommissioning.

Document Production Workflow

This set of topics describes the production flow of creating transaction documents, from data to doorstep, or host to post as it has been described. The key element here is the variable nature of the data: every document is unique, and a core facet of workflow management is ensuring the integrity of the production process at every step in the chain.

Data

Transaction documents generally represent a point in time snapshot of (typically a customers') data. This topic describes the types of data likely to be encountered; its extraction, normalization and formatting into human-readable form, and collation ready for the Composition step.

Document Objects

Document Objects are the myriad items used repeatedly across many documents and document types, both to present the variable data and as other components of the composed page and its layout. They include text objects, fonts, images, graphics, tables, and style sheets. This topic explains their intricacies and variations, and how to best manage them.

Composition

Composition systems play an extremely important role in document production. These systems apply business and document design rules to convert data and document objects into print files containing documents and mail-pieces. The topic discusses output file types (print, display, XML), integrity files, layout fundamentals, conditional processing, and postal preparation.

Print Streams

The print stream describes the document and carries the document contents to its presentation medium (paper or display), archive and other destinations. The topic describes the major print streams, their key attributes and their applications,

Transformations

The topic describes manipulations (changes to a print stream such as adding barcodes, making use of white space or carrying out address hygiene) and transformations (converting from one print stream type to another).

Print Management

Print management systems optimize the production workflow, maximize the effective use of available equipment through workload management, provide breakdowns of production costs, and ensure that all documents are printed.

Electronic Presentation

Transaction mail production includes the creation of legal documents that will be delivered electronically. The topic discusses the general principles of multi-channel / multi-media presentation, the options available, and when to use them. This includes PDF, XML schemas, and digital signatures. It also includes any technology or practices that prove the authenticity and irrefutability of the document.

Web and Mobile Delivery

Device characteristics, user experience and document access patterns differ widely across different sizes and forms of device. This topic discusses these facets of the document journey, and highlights considerations for successful implementations.

Archiving

As part of the production of transaction documents, organizations usually keep a legal copy for future reference. These copies-of-record service include customer service objectives (call center queries, requests for additional copies, court orders, etc.), and legal retention requirements. The topic discusses business reasons to archive, promoting system-generated correspondence into an archive, and indexing metadata.

Printing Technology

High-speed printing systems create most of the transaction documents within our industry. This topic describes the history of transaction document printers, and describes laser and inkjet printer features and characteristics, inks and toners, and cut sheet and continuous models.

Inserting Technology

Insertion (or enveloping) is the most mechanically complex, the least automated, and the most subject to environmental factors and operator error. The topic discusses the sub-components of a production inserter, use of barcodes to manage the process, software-based document control, and envelope ink-jet printing.

Delivery Process

This topic discusses postal history, treaties, regulations, processes, optimization of the cost of postage, postal barcodes (such as USPS Intelligent Mail Barcode), and envelope and address specifications.

So, having briefly summarized the steps in both the development life cycle and production workflow, the next sections of the EDBOK consider them all in much greater detail.

Document Systems Development Lifecycle

Requirements Gathering

Contributed by Paul Abdool edp

Knowing where you want to go (strategy) and what you need to do to get there (tactical steps) are critical elements to any project. Requirements Gathering covers the compilation and documenting of needs, necessities or obligatory elements prior to taking those strategic and tactical steps. It is a process.

In the Document Lifecycle, there are many players, agendas and requirements, so taking a panoramic look at an organization's needs prior to a project and incorporating those needs will lead to the best result – satisfied stakeholders, on-time execution and a good return on investment.

Establishing the Current State

Most industry pundits agree that "starting at the beginning" and following current procedures to the end of the process is important to complete a baseline assessment (Craine, 2000). This baseline is established by determining what processes and tools are currently in place. The current state should identify the following:

- technology in place,
- processes used,
- operational costs,
- utilization of assets,
- Key Performance Indicators (KPIs),
- legal and regulatory compliance elements, and
- strengths, issues and challenges.

This information is garnered through physical inventories, visual observation and a cross-section of stakeholder interviews. In some cases requirements gathering is an informal process and in others cases it is more formal. The formal model is usually more efficient as it mimics a military-like, top-down, accountability format.

Formal Model

The steps in the formal model are:

1. Establish a Project.
2. Hold a Core Team kick-off meeting to review roles.
3. Hold a Management expectation meeting.
    - Define business goals and high-level objectives.
    - Review the current state.
    - Establish risk and success criteria to ensure all parties are aware of the variables and potential hurdles.

4. Create a project plan to ensure the project stays on track.
5. Distribute a formal communication or an official project charter from a senior executive to ensure all stakeholders are aware of the desired outcomes and direction.
6. Hold stakeholder meetings to gather data and requirements.
7. Facilitate focus groups or user surveys to provide a wider range of opinions and increase buy-in.

The Current State is now established.

The formal model usually has set milestones to keep all parties accountable and to encourage participation.

Establishing a Project

Document/communication projects involve many business units. They are usually kicked off due to a problem or challenge. Problems may be related to operational efficiencies, mistakes or outside audits demonstrating deficiencies. Challenges may come from demands by management to cut costs, a reduction of staff or an increase in output requirements. Once one or more of these have been outlined by someone and they have been presented to management, a project is born.

In some cases management may stimulate this activity. They may choose to have the project lead by a well-rounded person with higher than average knowledge of the overall process and the various business units driving customer communications.

Core Team

Although this book is not about project management, Requirements Gathering requires organization and good documentation to:

- ensure completeness,
- report to project sponsors,
- acquire funding and support, and
- get the plan executed properly.

To get things started and organized the team size needs to be managed and everyone needs to know their role. The Project Stakeholders universe shown in Figure 1 demonstrates that many players are involved during the process. The *Core Team* needs to keep control and to make sure that the others are receiving the right amount of information through regular communication. This will keep the others engaged and up-to-date about the project.

In the initial Core Team kick-off meeting an outline of the goals and objectives should be established. To bring further structure to the process, the goals and objectives should be mapped to the calendar to align with the organization's known current projects. This prepares the team for the next step – the Management Expectation meeting.

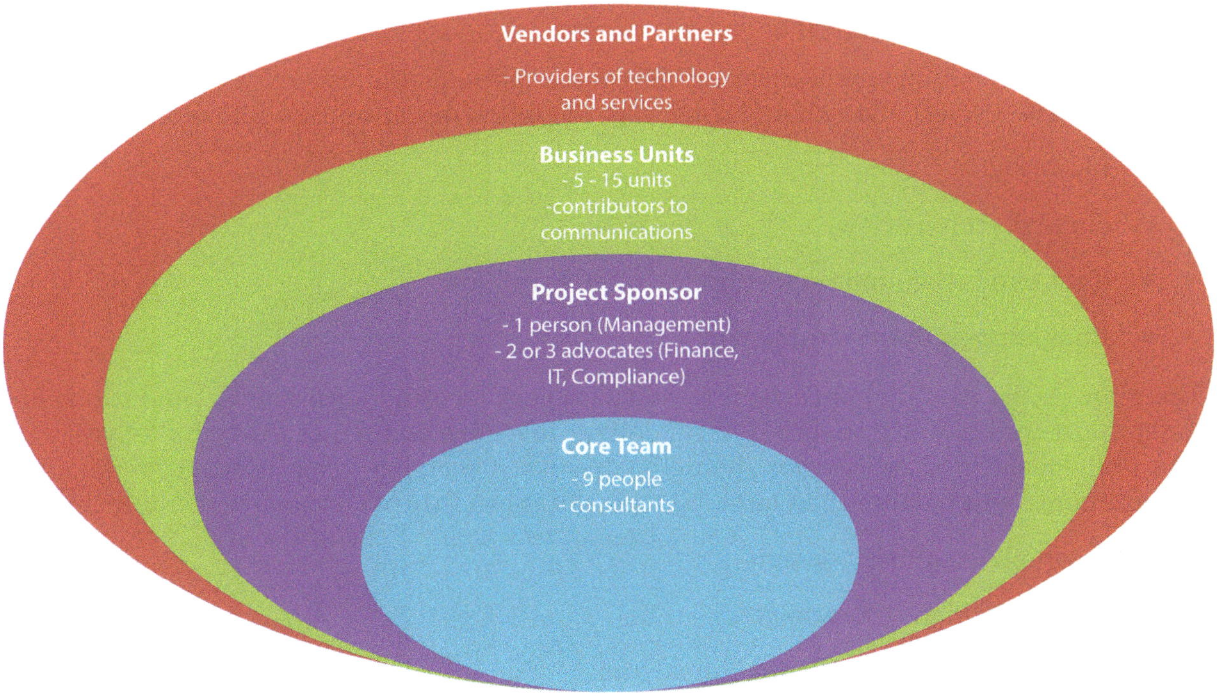

Figure 1 – Sample Project Stakeholders.

Management Expectation Meeting

Once the Core Team has established the goals and objectives, and created a high-level, calendared plan, it is time to present version one of the approach. A critical facet of the management meeting is the discussion of risk and success criteria to make sure all parties are aware of the variables and potential hurdles. Management will usually have more insight into current projects or projects that are on the horizon, which will allow the Core Team to ascertain the feasibility of getting the project rolling.

If the project is sponsored by management, then the management team will want to understand the approach and should support the navigation of the information gathering. They will act as liaisons to engage subject matter experts in different business units, which will expedite the project and create good communication.

Eventually management will want a more in-depth project plan and overall budget figures to ensure it can be executed with a solid return on investment or a reasonable payback period. The foundation of the Business Analysis will be created during the next few steps:

- detailed project plan,
- communication protocols,
- stakeholder meetings
- Subject Matter Expert (SME) interviews,
- focus groups, and
- surveys.

Project Plan Creation

After the initial management meeting the project plan should be revised and milestones adjusted. The Project Plan should consist of the following minimum milestones:

- planning,
- current state review,
- data validation and correction (as necessary), and
- final current state established.

The plan should also include a regular status report meeting with management to keep them up-to-speed and to ensure that all deliverables are being met. Any challenges that are jeopardizing milestones should be discussed because weak information about requirements will be amplified later. Samples of these future challenges could include:

- incomplete solution design,
- poor return on investments,
- longer implementation timelines, and
- compliance breaches.

Once the plan is set, a formal communication or an official project charter is distributed from a senior executive to ensure all stakeholders are aware of the desired outcomes and direction.

Gathering data and requirements is part art and part science. The project plan creates order and accountability. However, getting the information to solidify the strategy and creating buy-in and momentum along the way is an art form. By this point the general direction is known, but evidence is required to back up the theories.

There are *givens* and *variables*. *Givens* are things that you know and variables are things that are unknown, but can be solved for. Examples of *givens* could be:

- document volume produced,
- frequency of communication,
- physical inventories,
- expiring leases,
- software in use,
- software *sunsetting* (i.e. – version is at end-of-life),
- current budgets,
- a current shortfall in the operation (i.e. – currently at production capacity),
- a new regulation,
- retiring knowledge workers, and
- merger/acquisition.

Figure 2 – Current State Equation.

All of the *givens* in your equation should be data that is gathered and discussed. The impacts of these may not be known at the time of acquisition, but as discussions are had with stakeholders the picture will become clearer and the unknown variables will be discovered.

Another key step prior to speaking with stakeholders is to visually observe the operation in action ahead of time. Understanding the general document production workflow and the overall environment will provide a greater understanding of the current state. Both the good and bad will be seen, as well as how the personnel have adapted to the environment and their day-to-day challenges.

Stakeholder Meetings

A cross-section of stakeholder meetings should be held to gather data and requirements. After explaining what the project is about and the possible direction it may take, it is time to ask some questions. During these interviews many areas and issues of the business unit may be discussed and it is easy to lose focus, so it is good to be prepared with a set of questions that will guide the session. Questions should elicit both positive and negative discussions. Samples questions might include:

- How is the Business Unit affected by the project or document operations?
- Describe the current workflow or your part in it?
- Do you have any other projects on the go?
- What do you like about the current state?
- What do you dislike about the current state?
- Are there any financial and process challenges or issues with the current state?
- What would you like to see in the future state?

It is also important to limit the number of people participating in the interview to three people or less. This allows everyone to be heard while stimulating further conversation between interviewees to enrich the content. Ideas can be enriched by interactions and discussion. This is especially critical when trying to document old processes that have not been documented in the past. This *tribal knowledge* lives, but can be difficult to extract from one person.

The average attention span is about 45 minutes; therefore it may be necessary to break up the interview into two or three sessions. This also serves a few other purposes:

- As interviewees digest the conversation, they may come to the next session with more details.
- They may have to go in search of data that they can share in the next session.
- They may realize that there is another subject matter expert that will have better answers for you.
- There could be some processes that are already documented that can be brought next time.
- They may simply be more focused at another time when they are not distracted by another task.

There are many different stakeholders. Understanding each of their requirements and constraints will make it easier to develop a positive outcome for all.

Core Team

The Core Team is usually made up of no more than six people from Operations, Information Technology and Finance. The team may include two to three Subject Matter Experts and/or External Consultants.

The target ***Critical Outcome(s)***:

- Create a business case for senior management and/or project sponsor.
- Implement a new and improved solution to achieve the goals and objectives.

Project Sponsor

This should be a senior manager with budget, infrastructure or compliance responsibility. The project sponsor may also have two or three advocates, usually from their peer group (Finance, IT, Compliance). The role of the project sponsor is to:

- drive the project,
- provide stakeholder liaison, and
- change management.

The target ***Critical Outcome(s)***:

- increase profit,
- improve processes, and
- comply with new regulations.

Business Units

Contributors to communications or business processes involved with customer communication may be drawn from between five and fifteen business units, including Finance/Billing, Marketing, Legal, Compliance, Operations, and Sales. The role of the Business Units is to:

- articulate issues, challenges and desired outcomes,
- provide detailed information about their role in the process, and
- Business Unit change management.

The target ***Critical Outcome(s)***:

- increase profit, and
- improve processes

Vendors and Partners

Depending on the complexity of the project and integration requirements there may be no vendor interaction or the participation of a variety of vendors and partners. The role of this group is to:

- provide solution design as part of the team of providers of technology and services, and
- educate.

The target ***Critical Outcome(s)***:

- sales,
- improve processes, and
- implementation support.

Subject Matter Expert (SME) Interviews

Subject Matter Experts exist in many areas. You will find them in all of the stakeholder groups. These people are indispensable. They will provide key information from workflows, history, unwritten challenges and processes to integration possibilities. Typically, they are satisfied with the way things are as they have either built them or fixed them and have things working. It is critical that they understand the big picture and the direction that management wants to go. If their current situation can be improved by providing them with the tools, infrastructure, financial support or someone to hear their frustrations, they can be enablers to success.

On the other hand, there will be a few that would like to build very complex infrastructures that will not be economically feasible or may take too long to implement. Maintaining focus on data gathering and technical requirements to support the business analysis phase is vital.

SMEs are just that – experts. Do not waste their time with low level conversation. Come prepared to the interview. Interview others first to create a base of knowledge. Be prepared with facts and a list of unanswered questions from the previous interviews. Focus on their expertise and ensure that your questions keep referring back to the project at hand.

The best way to start is to state the givens than you know. Map the workflow and ask them to confirm or correct your drawing. If they are making multiple corrections that create confusion, ask them to map out the workflow for you and ensure that you understand it. Workflows should include at least the following elements:

- **Data / Inputs:** Source and owners of the data.
- **Composition / Design:** How are the documents are composed or created from the data?
- **Post-composition:** How are the documents are adjusted, transformed or prepared for output?
- **Archive:** How is the output stored, how long it is retained, when it will be purged?
- **Output and Distribution:** Electronic (emailed / web content) or physical (printed and mailed).
- **Tracking and Reporting:** How are documents tracked and productivity reported?

It is critical to keep detailed notes about the components of the workflow if even you have not heard of them or if they use a different vernacular specific to their organization. These will come up again. It also provides a list for further research with other SMEs or to provide details to vendors.

Facilitating Focus Groups

Depending on the project, testing concepts or soliciting feedback early on can help to find the holes in data or to add some credibility to the current direction. This voice of the internal or external customer or even SMEs may provide information that has not been thought of by people that are very close to the project. This feedback will provide a good indication of shortcomings and the future solution adoption success.

Professor Glenn Blank (Blank, 2002) of Lehigh University suggests that a focus group session should concentrate on:

- gathering opinions, beliefs, and attitudes about issues of interest to the organization,
- testing your assumptions,
- encouraging discussion about a particular topic,
- building excitement from spontaneous combination of participants' comments, and
- providing an opportunity to learn more about a topic or issue.

Focus group methodologies differ and are well documented. The key is to ensure that the subject is clear and that a good moderator facilitates the session involving all of the participants. Limit the group to no more than a dozen people to be effective for a session of 60-90 minutes. Here are sample steps from David Pitre (Pitre, 2010) of Davis & Company that you can use to prepare for a focus group session:

1. Determine your research question (or subject of focus).
2. Manage logistics.
3. Recruit and invite participants.
4. Develop questions.

5. Conduct the focus group(s).
6. Analyze your notes.
7. Summarize your findings.

User Surveys

Sometimes a wider audience is necessary to ensure awareness and to gain an understanding of the support for change. This panoramic view can be accommodated by online surveys or town hall type meetings. A skilled person in survey development is important to make sure that information is statistically sound or that the confidence in its accuracy is high. Sample size and test questions within the survey are important.

According to CustomInsight (CustomInsight, 2013), a provider of online HR assessment and development tools, determining the correct sample size requires three pieces of information:

1. The size of your population.
2. Your desired error level (e.g. 5%).
3. Your desired level of confidence (e.g. 95%).

For example, if you have a company population of 1,000 and you would like a confidence level of 95% with an error rate of 5% or range of +/- 5%, you would need 278 people to participate out of 695 if there was a 40% response rate. The calculator can be found on the following website:

http://www.custominsight.com/articles/random-sample-calculator.asp

Current State – Version One

After all data and requirements gathering is complete, the information should be presented in a logical manner for validation prior to the business analysis step. A samples ***Findings Report*** is shown in Figure 3.

Current State Review

When the Current State document is complete, it is time to gather the Project Sponsor, the Core Team and any key Stakeholders to review and validate the Findings Report. This step accomplishes a few things:

- It completes a key milestone; the review process brings stakeholders together once again to collaborate.
- It eliminates any misconceptions.
- It usually reveals some unknown facts.

Data Validation and Correction

Once the current state document is presented, it needs to be validated, adjusted, and/or corrected. This step is critical to ensure that everyone agrees on starting point. From here, solutions can be developed based on this snapshot in time. Without this step, solution development and business analysis activities do not have a foundation.

Figure 3 – Sample Current State document content.

This is not a new concept. In Eliyahu M. Goldratt's (Goldratt, 1984) book titled ***The Goal***, he introduced the Theory of Constraints (TOC), an overall management philosophy intended to help organizations attain their goals. Within the book, *Thinking Processes* were discussed. The order of operations in the Thinking Processes helps create a buy-in process just like the Management Expectation Meeting and Data Validation steps. The *Thinking Processes* are:

- gain agreement on the problem,
- gain agreement on the direction for a solution,
- gain agreement that the solution solves the problem,
- agree to overcome any potential negative ramifications, and
- agree to overcome any obstacles to implementation.

The Core Team and Project Sponsor must remain focused on completing this task since every elapsed day weakens the Current State as new variables are introduced by business units and economic pressures. The following should be established:

- deadlines for correction submission, and
- schedule the presentation or delivery of version two of the current state documentation.

The project sponsor must solidify the:

- business analysis plan,
- RFI/RFP development, and
- the team lead.

Informal Model

The steps in the informal model are similar to the formal model, but as the name indicates it is informal or more relaxed. The accountability, the scheduling, the timelines and the urgency are not present. This activity is usually led by a change agent to create a groundswell to enlist supporters. Informal models often evolve into formal models depending on the culture of the organization or the financial impact. The steps in the informal model are as follows:

- an individual or a Core Team kick-off meeting is held to establish the goal or to discuss the concept,
- stakeholder meetings are held on an *ad hoc* basis,
- management is informed on occasion of the work that is being done, and
- the current state is established with or without validation.

Identifying the Gaps and Business Problem(s)

As the data is gathered to create the current state, gaps in processes are identified, consistent requirements become apparent and business problems are found. Documenting the impact of the gaps and business problems during this stage helps to support the Business Analysis component. Gaps can include some of the following:

competencies or incompetence, performance levels, or internal and external customer impact.

During this stage, the project leader or consultant identifies business problems. Identifying the business problems can be challenging. Sometimes the symptom of a problem emerges to get the attention of stakeholders, but the root cause is not as easily identifiable. A simple analogy can clarify the concept. A weed appearing in the grass (above the surface) is the symptom or the effect of a root that germinated from a seed (under the surface). If the weed is cut or removed without the root then the problem weed will continue to surface.

The source of the problem can take many forms - dated methodologies and technologies, lack of budget and resources, or ignorance of upstream business requirements and downstream operational processes – so getting to the root of the problem and articulating the options, provides the key to the next steps – Business Analysis, Technical Analysis, Stakeholder Agreement, Design Specifications and Project Execution.

There are many ways to get to the specific underlying causes. These three techniques can stimulate your own approach. The three approaches are:

- The 5-Why Analysis
- Root Cause Analysis (RCA)
- Gap Analysis

The 5-Why Analysis[1]

Masaaki Imai (Imai, 1986) founded the Kaizen Institute Consulting Group (KICG) to help western companies introduce the concepts, systems and tools of Kaizen[2]. Since then it has been used in various permutations throughout the world. Karn Bulsuk (Bulsuk, 2009) states that "the 5-Why Analysis is used throughout the kaizen concept and in quality control, is a tool to discover the root causes of a problem." The methodology truly follows the name and suggests that the interviewer ask the interviewee 'why' up to five times to get to the root cause.

1. Identify the problem.
2. Ask yourself: why did this happen? Come up with all the causes you can think of.
3. For each of the causes you just identified, ask "why did this happen?" again.
4. Repeat until you've done steps 2 and 3 five times. You should have identified the root cause by this stage.
5. Find solutions and countermeasures to fix the root cause.

1 Root Cause Analysis (RCA) – RCA is a tool designed to help identify not only what and how an event occurred, but also why it happened.

2 Japanese for "improvement" or "change for the best" refers to philosophy or practices that focus upon continuous improvement of processes in manufacturing, engineering, and business management.

Root Cause Analysis (RCA)

In ***Root Cause Analysis for Beginners*** by James J. Rooney and Lee N. Vanden Heuvel (Rooney & Vanden Heuvel, 2004), they state that "RCA helps identify what, how and why something happened, thus preventing recurrence." Their process involves the following four major steps:

1. Data collection.
2. Cause charting.
3. Root cause identification.
4. Recommendation generation and implementation.

Data collection has been discussed throughout this section. The key point that Rooney and Vanden Heuvel make is that data collection takes the most time, but it is absolutely necessary because root causes cannot be identified without having complete information prior to analysis.

Their *Causal Factor Charting* step "provides a structure for investigators to organize and analyze the information gathered during the investigation and identify gaps and deficiencies in knowledge as the investigation progresses." This is very similar to the SME workflow whiteboard session. The difference is that the causal factor chart diagram includes logic tests that describe the events leading up to an occurrence, plus the conditions surrounding these events.

The third step is root cause identification.This step entails "the use of a decision diagram called the Root Cause Map to identify the underlying reason or reasons for each causal factor." The map, much like the SME workflow diagrams, creates a starting point for the interviewer to get questions answered about causal factors and why happen.

The final step of recommendation generation and implementation crosses into business analysis.

Gap Analysis

Gap Analysis is a technique that people use to determine what steps need to be taken in order to move a business or project from its current state to its desired, future state (Business Dictionary, 2014). Once the current state is documented and as the future state becomes defined more clearly after further business analysis and solution design, the gaps become evident. The gap between the two states creates requirements to close the delta.

Insight from finance, IT, vendors and consultants can help to solidify the analysis and lead to a sound business decision. A few factors to consider at this stage are:

- financial forecasting and planning,
- regulations and applicable laws,
- suppliers,
- staff skills,

- economic conditions,
- opportunity costs,
- competition, and
- available technology.

Alignment of Leadership Vision and Operations

Most of the time, leaders have a vision of what they would like to occur. Just as often, they are unaware of what options exist to achieve the vision or the implications on their Operations team. The vision usually has a simple objective – improve the bottom line – make money or save money. Operations are usually tasked with saving money through efficiency while the business units are trying to both, improve the top line, while reducing expenses if possible. Aligning these two or making the rubber hit the road is part art and science.

References

Business Dictionary. (2014). Retrieved from http://www.businessdictionary.com

Blank, G. (2002). Professor. Bethlehem, PA, USA: Lehigh University.

Bulsuk, K. G. (2009). An Introduction to 5-why. Retrieved from KarnGBulsuk: http://www.bulsuk.com/2009/03/5-why-finding-root-causes.html

Craine, K. (2000). Designing a Document Strategy. MC2 Books.

CustomInsight. (2013). Retrieved from Custom Insight: http://www.custominsight.com/

Goldratt, E. M. (1984). The Goal.

Imai, M. (1986). Kaizen: The key to Japan's competitive success. New York, NY: McGraw-Hill.

Merriam-Webster, I. (2013). Merriam-Webster Online Dictionary.

Pitre, D. (2010). How to Conduct Employee Focus Groups - fast! Glenrock, NJ, USA.

Rooney, J. J., & Vanden Heuvel, L. N. (2004). ASQ (American Society of Quality Control). Retrieved from http://asq.org

WebFinance, I. (2013). BusinessDictionary.com

Business Analysis

Contributed by Paul Abdool edp

What is Business Analysis?

The International Institute of Business Analysis (IIBA) defines Business Analysis as a set of tasks and techniques used to work as a liaison among stakeholders to understand the structure, policies, and operations of an organization and recommend solutions that enable the organization to achieve its goals. (***What is Business Analysis?***, 2014)

Business Analysis Inputs

As you can see, this definition is long and incorporates many areas. In essence, Business Analysis should not be conducted in a vacuum. Although there can be many analyses done in parallel to create the business case and overall solution, it is critical that a panoramic view of multiple factors is considered. In Figure 1 there are examples of factors that may influence a business decision.

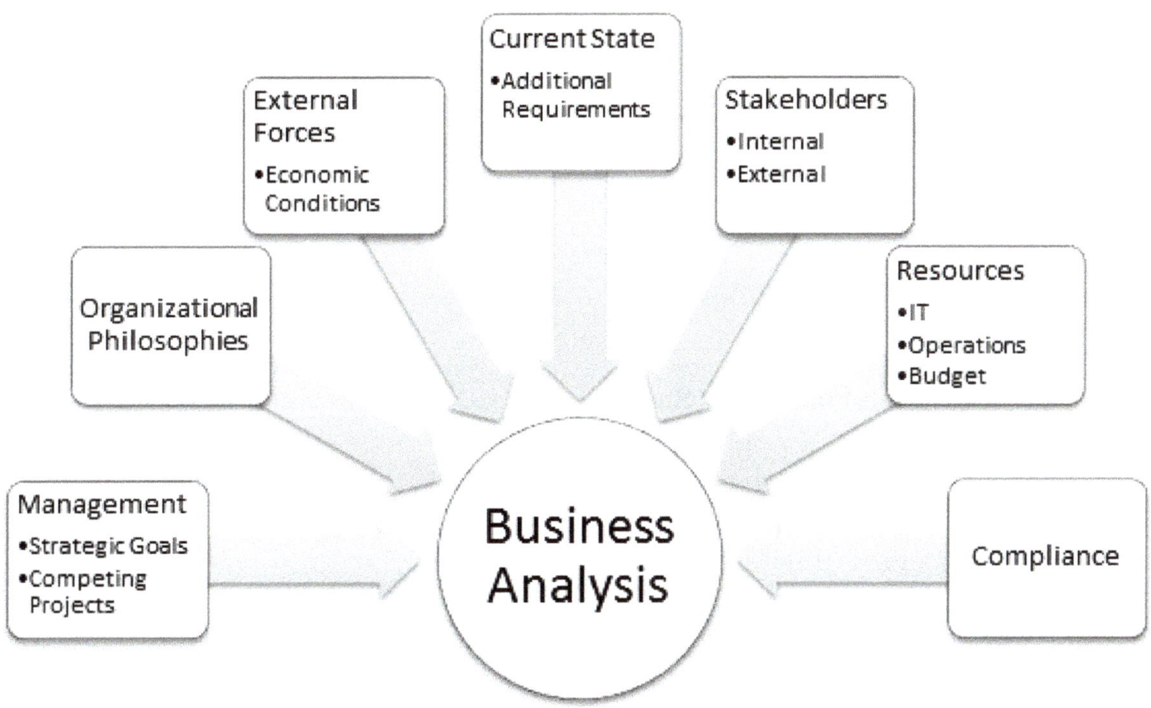

Figure 1 - Business Analysis Inputs.

Business Analysis Outputs

The Business Analysis should culminate with a Business Case complete with some or all of the following components:

- Executive Summary
- Current State
    - Methodology
    - Major Issues and Risks
    - Inventory of Software and Hardware
    - Data Source List
    - Applications List
    - Workflow Diagrams
    - Interview Summaries
- Recommendations
    - List of Improvements
    - Other Qualitative Benefits
- Financials
    - Cost-Benefit Analysis
    - Return on Investment
    - Payback Period Details
    - Opportunity Cost Information
    - Risk Descriptions
- Implementation Plan / Next Steps
    - Key Performance Indicator Monitoring
- Appendix
    - Extra Details (if necessary)

Business Analysts

The people who carry out business analysis tasks are called business analysts or BAs. Business Analysts take all forms.

Internal

Some Business Analysts are generalists that support multiple areas within their organizations, while others are focused on serving specific business units or internal customers. When a BA becomes more specialized, their particular discipline may precede the BA title. For example, if they work in Information Technology they may be referred to as IT Business Analysts or Systems Business Analysts.

Internal BAs may need external support as projects enter arenas where niche knowledge and experience is required.

External

Industry Analysts – Fee-based

Industry Analysts provide project-based services designed to assist clients with technology selection and business process decisions. They research and survey the entire industry or segments of it to predict trends that will impact businesses. A sample of industry analysts within the electronic document arena are: the Gartner Group, Forrester Research, InfoTrends, IDC, Caslon, and Madison Advisors.

Consultants – Fee-based

Some of the organizations mentioned above offer independent consulting, however there are many customer communication consultants that can support organizations during business case development. Consultants can be independent and may also possess certain specialties within the document lifecycle or within an operation. Using consultants can prove to be beneficial since they focus on creating business cases every day, and are aware of trends and recent innovations. They also have proven methodologies and experience so your organization does not have to reinvent the wheel.

Vendors – Fee-based or Complimentary (Cost of Sales)

By nature, vendors must create business cases every day to make sales. Their services may or may not be fee-based. If their consulting is complimentary, then their business case may include their service or product offerings. Vendors also provide a lot of educational seminars and events that can help to round out your business analyst's knowledge prior to getting a project going.

Business Case Success Elements

Business case success requires three major elements to get a project moving. They are:

- senior management support,
- impactful business arguments, and
- stakeholder involvement.

Do not underestimate the importance of these elements. They can make or break a project.

The number one reason projects do not move forward is because senior management[1] was not involved from the beginning. When senior management agrees that a problem or challenge exists and assigns resources to bring a solution forward a project starts down the right path.

The second hurdle that can slow project momentum is a weak *business case* presented to the supporting senior management team. Although a business case may be good in individual departments or may solve a problem, the overall impact to an organization may be looked at as uneventful, like a mosquito bite on your back. The bite may bother

1 Individuals that have responsibility for financial budgets and who possess the authority to make internal changes and sign legal contracts.

you, but nobody else is affected. So, it is important to articulate the significance of the solution options to the people who are responsible for impacting the organization. It is also critical to describe what will happen if a solution is not implemented. For example, if the mosquito carries an infectious disease and the mosquito population is growing.

Another reason projects do not move forward is because the stakeholder involvement was not inclusive enough. If a business case does not consider the ramifications that the solution may have on stakeholders outside of the immediate sphere, stakeholders can feel like their areas do not matter. This can create obstacles to project approval or access to the necessary resources required to implement a solution. In other words, if a business case is created by Operations to acquire software, but they need IT to implement it, they should involve IT from the beginning. Technical Analysis is not mentioned here, but both internal and external Subject Matter Experts can help to ensure that the solution options in the business case are technically feasible and sound.

Business Analysis Players

In the electronic document field, business analysts must ensure that many players are involved during project or business case development. They must participate in information sharing, solutions validation, and ensuring that the financial analysis is thorough and complete. Here are two lists: Stakeholders and Participants.

Typical Stakeholders

- Business Process Owners
- Information Technology (IT)
- Finance
- Compliance
- Legal
- Marketing
- Customer Service
- Data Center / Print Operations
- Mail Center

Participants

- Project Manager
- Operations
- Finance
- Compliance
- Quality Assurance
- Print Operations
- Business Systems Analysts

- Internal and External Subject Matter Experts (SMEs)
- Vendors / Suppliers

Value of Business Analysis

Stakeholder Consideration

In the electronic document domain, business analysis must take into consideration the various stakeholder interactions and correlations. Figure 2 shows an example of dynamic customer communication creation and ongoing adjustments. Concepts may develop in the Marketing Department based on data analytics or a financial goal from the Finance Department, but a customer communication cannot be prepared properly without the right tools in the Information Technology and Operations departments. In addition, all of those departments must be cognizant of the Compliance department's requirements and other concerns such as postal systems. Of course, no organization can operate without customers.

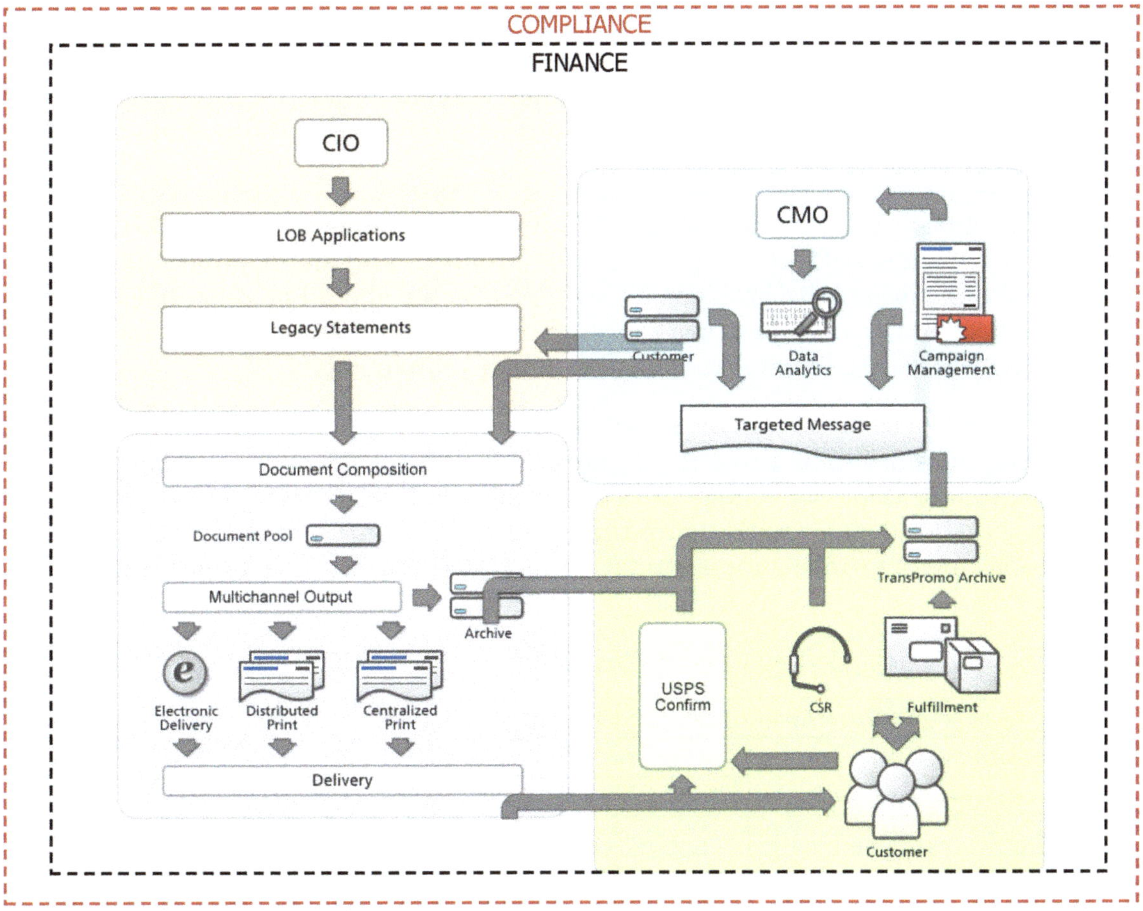

Figure 2 - Sample Stakeholder Interaction.

Business Case Development

"A business case is an argument, usually documented, that is intended to convince a decision maker to approve some kind of action. The document itself is usually referred to as a business case." (Rouse, 2012)

Business case development involves a detailed analysis of the Current State to evolve to the Future State. This investigation into what will be involved in producing the deliverables of a project is critical for getting it from started to solution implemented.

Key outcomes from a good business case and analysis are:

- a firm scope,
- a comprehensive, documented set of requirements,
- a cost-benefit analysis, and
- a set of next steps.

Getting Project Approval

Getting project approval once a business case is developed is the final step. The business case is critical because no one can make a decision without all of the facts. The business case must be easy to understand and must utilize the nomenclature of the organization, with a limited amount of industry terminology so it is not misconstrued. The challenges, reasons or pains involved must be described in both financial and other business impact terms, such as compliance, human resources, customer service, reputation and brand integrity. Then the solution must counter those challenges with the positives outweighing the negatives.

The sample solution table below consists of a problem statement, a description of the problem, the solution/product used to solve the problem, and an explanation of how the solution will counter the problems. This type of table should be repeated for each problem as necessary. It will ensure thoroughness and force the project team to be prepared for presenting the business case to senior management

During business case development, solution options are tested to determine if they are feasible. The testing can be through presentations or discussions with subject matter experts to validate that cost components and implementation hurdles have not been left out or overstated. This will ensure that the overall solution includes a "rough order of magnitude" for the senior management team to understand the impact of their final decisions.

Governance, Risk Management and Compliance

Increased Production Demand (Problem Statement)		
Problem Description	**Solution/Product**	**Explanation**
Overutilized mailpiece inserting resources	(2) Production Inserters	Increase the amount of mailpieces insersted automatically
Productivity		Inserts at 10,000+ pages per hour
Capacity		Will handle expected growth; 1,200,000 per year
Document Integrity (OMR)		2D barcode capability
Reports		Standard productivity reporting
Redundancy		With 2 inserters
In-line Postage Meters		Reduce extra step and costs of offline postage application
Load Balancing		Split work across both inserters to increase inegrity and capacity
Increasing Production Costs		Reduce costs by 25% or by $150,000/year due to reduced manual insertion

Table 1 – Sample Problem-Solution Table.

Over the past few years, governance, risk management and compliance departments have been appearing in many organizations. Business case development requires discussion and analysis of these areas to ensure that procedures are aligned and proposed solutions are addressing any regulatory requirements. This allows senior management to assess both the qualitative and quantitative risks to the organization. By incorporating these in the business case it will reduce questions and doubts about the feasibility of the proposed solutions. In addition, by involving these groups early in the process, it will ensure that the business case will not be derailed because of missed information.

Establishing Key Performance Indicators (KPIs)

One of the requirements that organizations have once the business analysis is complete is a plan to continually measure how things have improved or changed after an implementation plan is in place. By establishing KPIs during the business analysis or blending them into the requirements of the solution the organization is compelled to maintain checks and balances to measure the success of the changes. Usually the KPIs have been established by the organization, but in some cases they have not been benchmarked. The current state of the KPIs should have been measured during the requirements gathering stage. Sample KPIs are:

- production throughput of document production lines (documents completed),
- Service Level Agreements (SLA) performance (met or not met),
- number of incidents or personal information breaches (more or less),
- document integrity levels – error count (more or less),
- cost and time to adjust or reprogram documents (more or less),
- overtime hours used (more or less), and
- internal customer satisfaction (complaint or compliment count).

Although this book is not about project management, project managers are critical during implementation and post-implementation reporting. They help to manage the scope, to keep the implementation on time and to manage the budget.

Business Continuity

More and more, organizations are being mandated to have business continuity plans. When a business case is established it should consider any business continuity plans and policies that an organization already has in place to make certain that it is in compliance. In the critical customer communication space there is no room for error or downtime.

Techniques

Cost-Benefit Analysis (CBA)

If the business analysis shows that the benefits do not outweigh the costs, the business case is not going to help any organization.

Most organizations usually have a cost-benefit analysis process for making decisions. Cost-Benefit Analysis has qualitative and quantitative objectives that answer the

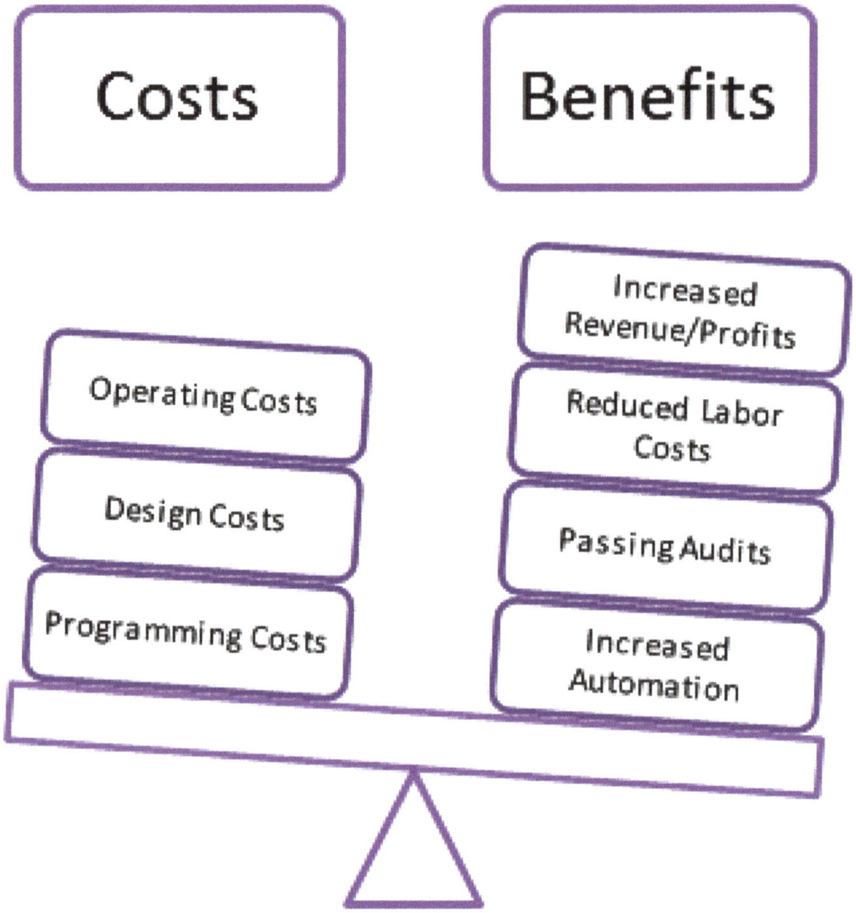

Figure 3 - Cost-Benefit Scale.

question, "Will the solution costs impact the organization in a positive way?" To determine the equation, data collected during requirements gathering and solution development are broken into the following categories:

1. Benefits (reduce costs, increase revenue).
2. Costs (increase costs, decrease revenue).

Figure 4 – Simple Cost-Benefit Equation.

If the project changes, by eliminating or incurring a benefit or cost, whether it is a one-time or recurring, business or technical, that benefit or cost should be included in the CBA.

Benefits

Benefits reduce costs or increase revenue. Management that is concerned about financial elements of the business analysis will want to hear about the elements that impact the bottom line by reducing costs or increasing revenue. Some are easier to measure than others. In most cases, Financial Officers are most concerned about reducing costs or avoiding anticipated costs (also known as cost avoidance). These are hard costs that are visible and predictable. Increasing revenue has many variables and is dependent on multiple factors and although possible to predict it is easier to refute.

Here is a list of sample benefits. In many cases they are aligned with an organization's KPIs:

- reduced operating costs:
  - reduced Labor Costs / less overtime hours used,
  - lower maintenance costs,
  - reduced power consumption,
  - increased Automation (increased throughput on document production lines),
  - lower inbound Call Center costs, and
  - lower postage / distribution costs.
- passing audits / less fines,
- Service Level Agreements (SLA) performance improvement,
- increased document integrity levels – error count lower, and
- higher customer satisfaction levels.

Costs

Costs are classified as elements that increase costs or decrease revenue.

- increasing design costs,
- growing programming costs,
- higher decommissioning costs,
- escalating call centre costs,
- rising sales and marketing costs,
- mounting consumables costs,
- decommissioning costs, and
- document, hardware or software design costs.

Another cost that is often overlooked, because it not often accounted for by Financial Officers on a balance sheet, is the Opportunity Cost or Cost of Inertia. This is usually shown separately as organizations deal with it differently.

Return on Investment (ROI)

Return on investment, or ROI, is the most common profitability ratio. It is a performance measure used to evaluate the efficiency of an investment or to compare the efficiency of a number of different investments. ROI evaluates the money you invest versus the return you realize on that money.

To calculate ROI, the benefit (return) of an investment is divided by the cost of the investment; the result is expressed as a percentage or a ratio (Investopedia US, A Division of IAC, 2014).

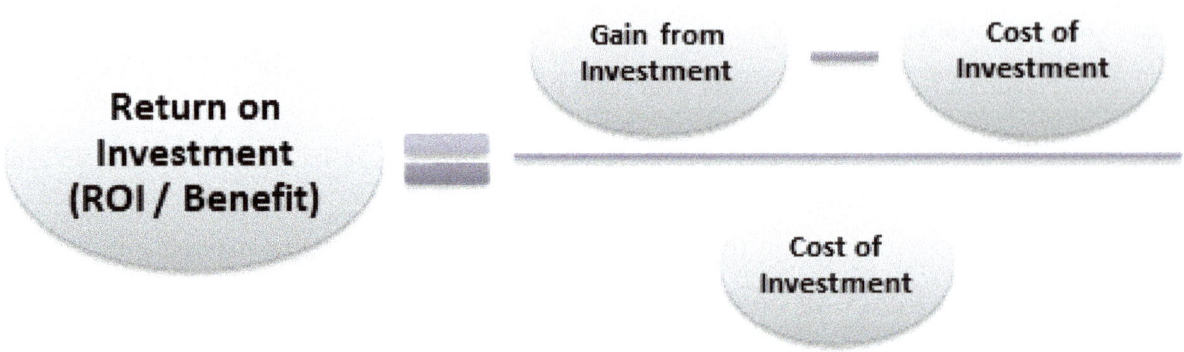

Figure 5 - ROI Equation.

If an investment does not have a positive ROI then the investment should be not be undertaken.

In the formula above, "gains from investment" refers to the amount of money saved or made from making the investment ("cost of investment"). An organization can use ROI in several ways to gauge the profitability of a particular direction a business decision will go. or example, the formula could be used to measure the effect of a change in inventory volume carried, fees paid and hardware and software investments.

Sample return on investment (ROI) equation:

Inputs:

- Cost of Investment and Implementation = $240,000
- Gain from Investment (cost savings) = $600,000 ($10,000 per month over 60 months)

Output:

- ROI = 150% or a net return of $360,000
  - 150% ROI over 60 month (5 years) or 30% per year

Payback Period

The point in time (usually in months) in which the initial investment is paid back.

Inputs:

- Cost of Investment and Implementation = $240,000
- Current State Cost - $400,000
- Future State Cost - $390,000
- Gain from Investment (cost savings) = $10,000 per month ($400,000 - $390,000)

Output:

- Payback period = 24 months
    - $240,000 / $10,000 = 24 months

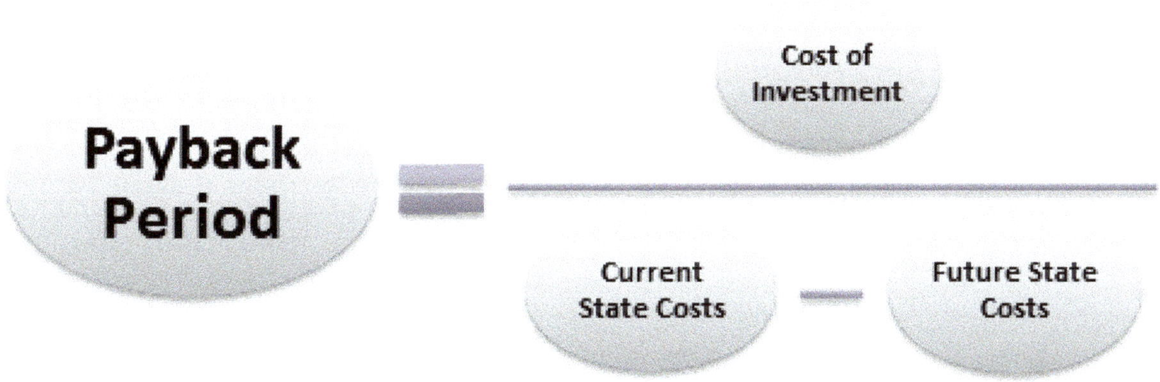

Figure 6 – Payback Period Equation.

PRESENT VALUE (PV)

Financial Officers often use the Present Value of the savings or gain from an investment as the Litmus Test for business cases. It is considered a more realistic view because it includes the value/cost of money. The current worth (Present Value - PV) of a future sum of money (C) or stream of cash flows given a specified rate of return (i) and factoring in the number of time periods (n) is the formula used below.

$$PV = \frac{C}{(1 + i)^{n}}$$

$470,653.69 = $600,000 / (1 + 0.10) by a factor of 5 years

Figure 7 - Present Value Equation.

In some cases, business analysis can be less formal and decisions can be made using other techniques, such as MOST or SWOT Analyses. These techniques are usually used when the business decision is smaller or the project size requires fewer resources (time, budget, management involvement) and has less risk associated with it. These techniques are also used to align an organizations strategic direction, which will help when developing Requests for Information, Quotations and Pricing (RFIs, RFQs or RFPs).

MOST

"This is used to perform an internal environmental analysis by defining the attributes of MOST to ensure that the project you are working on is aligned to each of the four attributes." (Strategy Consulting Limited, 2013).

The four attributes of MOST:

- Mission (where the business intends to go),
- Objectives (the key goals which will help achieve the mission),
- Strategies (options for moving forward), and
- Tactics (how strategies are put into action).

SWOT

This is used to help focus activities into areas of strength and where the greatest opportunities lie. This is used to identify the dangers that take the form of weaknesses, and both internal and external threats. (Humphrey, 2005)

The four attributes of SWOT analysis:

- Strengths
  - What are the advantages?
  - What is currently done well?
- Weaknesses
  - What could be improved?
  - What is done poorly?
- Opportunities
  - What good opportunities face the organization or business unit?
  - Where are your competitors performing unsuccessfully?
- Threats
  - What obstacles does the organization or business unit face?
  - Where are your competitors performing well?

Next Steps and/or Implementation Plans

Once the business analysis is complete, the final step to finish the business case is to develop a logical and sequential set of next steps or an implementation plan. If senior management likes the business analysis component they will ask, "Now what do we do?"

These next steps should be laid out in chronological order with the associated project resources and financial investments aligned within the timeline. Most organizations have a project management team that will be able to support these efforts. Some organizations have a Change Management team that can also supportive. A sample high-level implementation plan is shown opposite.

Timing	Actions
2011-Q3	Executive presentation (Information Security and Privacy Steering Committee)
	Project approval
	Order 2 New Printers (lease)
	Start Change Management process
	Technical Statement of Work (SOW) creation and Project Plan development
	Install Server and Software
	Software Training for 2 Operations Staff members
	Add Job Ticket to Print Output
	Indexing and creating of the piece level workflow and setup job types for each job
	Establish SLA Monitoring
	Implement Job level Tracking
	Enhance OMR marks
	Install New Cameras on 3 inserters
	Add 2D barcodes and add sequence numbers to all documents running through Mail Services
	Install New Printers
	Use all old envelope stock
	Redesign all envelopes to be outside side seam
	Order new envelopes with Permits
	Staff reductions; Print Operator
	Staff reductions; Interoffice Associate
	Begin Scanning Process review
	Put KPIs in place and begin automated measurement and reporting
2011-Q4	Migrate all checks to cut sheerr printers
	Install Postal Sorting Software
	Build address data from data source and format for integration into Postal Software
	Develop Postal Document Requirements and obtain approval from Postal Service
	Begin Migration from Metering to Permits on envelopes
	Eliminate meters
	Retire Sorter
	Staff reductions; Return mail
	Staff reductions; Inserter Operator
	Staff reductions; Outgoing Mail / Sorting Lead
	Establish Document Revitalization Order - Base on Volumes and Risk factors
	Begin Return mail review
	Do research with marketing and Compliance to see what docoumnts could be insourced
2012-Q1	Go live
2012-Q2	Establish new goals based on data captured in 2012

Figure 8 - Sample Implementation Plan.

Key Performance Indicator (KPIs) Monitoring

Business Cases should include a way to continually check on what will be implemented. With today's technology KPI Monitoring should be fairly automated. Most software systems, hardware systems, and line of business applications have the ability to create reports in some way. Most of the time management want reports in a certain format, so ensuring that systems have an application programming interface (API) can make this easier.

The implementation plan or long-term project plan should include milestones or check points to revisit the success or failure of components of the project. This is critical for budgeting, financial reports, and for future funding.

Two critical elements that are prevalent today are savings reports and compliance or audit reports. Savings Reports should be cumulated to demonstrate the ROI or Payback Period to Financial Officers and management. If the business analysis was done correctly and conservatively, these reports should exceed expectations. If they do not, it is important to understand why and make adjustments. In some cases, the original senior officer may need to be informed of the areas that are creating challenges so corrective measures can be taken or the savings expectations need to be adjusted accordingly.

Compliance or audit reporting can be done internally or externally. They are great "tests" of the success of the program or where the weak links are. If the business analysis was done correctly, the KPIs should be aligned with these periodic reports and the core team or participants should be able to make adjustments between audits to ensure compliance or improvements.

References

Humphrey, A. (2005). SWOT Analysis for Management Consulting. SRI Alumni Newsletter.

Investopedia US, A Division of IAC. (2014). Profitability Ratio. Retrieved February 2014, from Investopedia: http://www.investopedia.com/terms/r/returnoninvestment.asp

Rouse, M. (2012, January). Business Terms Glossary. (M. Rouse, Ed.) Retrieved February 2014, from WhatIs.com.

Strategy Consulting Limited. (2013). exploring corporate strategy using m.o.s.t. analysis. Retrieved February 2014, from http://www.strategyconsultinglimited.co.uk: http://www.strategyconsultinglimited.co.uk/business-strategy/exploring-corporate-strategy-using-m-o-s-t-analysis/

The Atlantic Systems Guild Limited. (n.d.). Mastering Business Analysis. Retrieved February 2014, from Volare Requirements Resources: http://www.volere.co.uk/mba.htm

What is Business Analysis? (2014, February). Retrieved 2014, from International Institute of Business Analysis (IIBA): http://www.iiba.org/babok-guide/babok-guide-online/chapter-one-introduction/1-2-what-is-business-analysis.aspx

Technical Analysis

Contributed by Stephen Poe edp

Why Do a Technical Analysis?

A properly executed technical analysis insures that all business requirements have been extracted from the customer and understood by the team. Those requirements should be reviewed with, prioritized, and approved by the customer, and then distilled into a feasible technical architecture and design for implementation. The technical analysis can then be referenced for direction and clarification when implementing the project and used by the customer to verify and validate that the final delivered project meets the requirements as agreed upon by both parties.

A technical analysis is a vital part of a successful business project. It is critical in insuring the project is successful, meets its goals and objectives, and reduces and mitigates whatever risks are inherent with the project.

Technical Analysis Processes

A typical technical analysis (TA) process is iterative and involves close interaction between both the project team and the customers for the project. The customer in this case can be either internal or external. Iteration is necessary to insure that the requirements and conclusions are correct, understood, and accepted by all parties. The actual process used will vary between organizations and based on the type of project being developed. The TA process for a new or refurbished plant for mailing operations will be different than one for developing a new Line of Business application and different again from one for a new product to be developed and released by a vendor. The next sections list some general guidelines. The term project is used here generically.

Vet the Project Business Definition

Normally a technical analysis starts with a project already defined at the business level. This business definition should include the business goals and the business constraints for the project. This typically includes the budget, resource, and timeline constraints to be vetted as part of the technical analysis, and any flexibility that might exist.

Ideally the business definition should state what needs to be done, not how it should be done. If a business definition contains too many hows, it should be viewed with suspicion. The TA team should go back to the business team and probe further. How definitions often mask the real pain point and can narrow the TA teams focus. Did the business definition really uncover the business needs? Sometimes in specifying a how too early in the process other potential solutions or options are overlooked.

Once a good business project definition is in hand, the TA can proceed.

The TA team needs to familiarize itself with the business definitions and make sure they understand them and the framework within which they set. If any of the business components are unclear they must be reviewed and clarified.

Review the Current Process

The technical analysis can now begin. If this is an upgrade or a replacement for an existing business project the TA team must review the current implementation, both overall and with an extra focus on current limitations and pain points. Although the current process should not necessarily be used as a starting point for the new process, an in-depth understanding of its limitations can be useful when formulating the new process. Whenever possible, interview the current users. Read the current documentation. See where the current user actions are different than what is defined in the documentation. Watch the current users perform their jobs. Frequently what a user says in an interview is different than what they actually do. An ethnographic study is ideal if you have UX (User Experience) experts available and it's relevant to the type of project. Ask the users what they would prefer to have. Initially make these questions open-ended. Do not suggest potential options; but let the users suggest them.

Future Requirements

The TA team should consider the future direction of the business and of the project within the business within a reasonable time frame, usually 2-3 years. The main purpose of this exercise is to ensure that technical decisions being made during the Technical Analysis won't inhibit the direction of the business or project going forward, will not cause extra work, and will not otherwise make the current project's work unusable due to projected future directions.

Non-Functional Requirements

Define the non-functional requirements (NFR). NFRs are requirements that, although directly related to the functionality the process must meet, are part of the environment in which the process lives. For example, if software, what is the current IT environment within which the final project has to operate? If on an operations floor, what physical environment and limitations are present?

Review the users. What skill sets do they have? What skill sets are they using, and what additional skill sets might they possess?

SLAs, KPIs, and other Metrics

Determine what SLAs (Service Level Agreements), KPIs (Key Performance Indicators), or equivalents are in place. These are usually hard requirements, frequently contractual, with specific numeric metrics that must be met. Examples include the number of documents processed per hour, number of mail trays produced in an 8-hour window, percentage of acceptable rework, maximum error rate, and any other metrics that govern the throughput, frequency, and quality of the final process. For each of these metrics, determine at least what the projected 2-3 year growth is, and, if possible, what the projected 5-year growth is. This insures your design and deliverable do not become obsolete 12 months after you deliver.

Wikipedia has a good introduction to Key Performance Indicators; see http://en.wikipedia.org/wiki/Performance_indicator.

Functional Requirements

The team should produce functional requirements. These are the actual requirements against which the project implementation will be developed. Functional requirements should be:

- **Comprehensive.** They should cover all of the functionality the project or process must deliver. Functionality that is obvious or already in place in the existing system should be included; nothing should be left out.
- **Granular.** Each functional requirement should cover one function at a low enough level that it can be tested for, but not at so low a level that it begins to dictate how the requirement should be implemented.
- **Clear.** The team developing the technical analysis is usually not the team implementing the project or process. If the requirements are not clear or are subject to interpretation, what gets implemented may not be what was originally intended.

There will also likely be requirements that will not be automated, but will be handled by way of new or changed manual operations or procedures for the end users. These requirements must be captured and documented, and passed on for design and development by the relevant group.

There should also be a section of known non-requirements. These are issues that have been discussed and will not be in the project (or, at least not in the first version). This section can eliminate the same questions being raised multiple times ("but have you considered x?") as well as providing the basis of a roadmap for future releases or upgrades.

For more background on functional specifications, see ***Painless Functional Specifications*** by Joel on Software at http://www.joelonsoftware.com/articles/fog0000000036.html.

Functional Requirement Formats

It is often helpful to develop user stories around the requirements. Instead of calling out a function like archival for instance, saying the user should be able to send a PDF copy of the document to the archive system of record via a drop down menu can help both the business users and TA team gain a clear understanding of the requirement.

One frequently used format for functional requirements is the SCRUM Theme/Epic/User Story format.

See, for example, Mike Cohn's discussion of ***User Stories, Epics and Themes*** at http://www.mountaingoatsoftware.com/blog/stories-epics-and-themes for an introduction to this best-practices functional requirement format.

Investigation

At times, requirements may need some research before they are presented to the end user. This is usually to determine the feasibility of a new function or an order of magnitude estimate as to how much work is required to deliver a function. This can help the business users prioritize and rationalize the desired functionality.

Functional Requirements Vetting

Ideally, functional requirements should be vetted with the end users. This may be difficult because there is little to nothing to show them. Two approaches are recommended; both should be used.

Initially, review each functional requirement with the end user community and have them confirm each one. If an end user cannot understand the functional requirement then it is not clear and must be rewritten. Frequently an end user requests a change: "I know that what I originally said we need, but now I really see that we need this instead." This should not be viewed as a failure on either side; it is an iterative process and may require several iterations before a final set of vetted functional requirements can be agreed upon. Do not move forward until everyone agrees on the final set.

Another critical component during this review process is prioritization. If you ask a user, "Would you like this feature?" they will say, "Yes." This creates feature bloat and a design that cannot be implemented within the constraints of the project. Instead, have users prioritize features: "Out of this list, what are your top three?" or, "For each requirement on this list, rate it as a 1 (highest priority) to a 3 (lowest priority)." It can also be helpful to let users know the relative cost of a feature: "This feature will take a lot of time and money and we won't be able to do a lot of others. Is it that important to you?" Another methodology is the MoSCoW method ('Must have', 'Should have', 'Could have', and 'Won't have'). Refer to http://en.wikipedia.org/wiki/MoSCoW_method for more information.

Some tools that can be included are wireframes, paper mock-ups, and physical mock-ups. These provide a better visual representation of what the TA team is proposing. The end users should be told that these are only initial mock-ups; the final delivered process of project will most likely not look the same.

Research

At this point frequently the team will need to identify and research issues, process, products, or technologies that might be necessary for the project. There may be new software, new hardware, emerging technologies, or new processes or methodologies available that the team should review and become familiar with before making decisions as to whether or not they might be appropriate as a component or basis for the new project.

Statement of Work

If the project is green-lighted to move forward, the TA team develops and delivers as its final report a Statement of Work. The SOW should contain (as appropriate):

- business project definition and assumptions,
- functional requirements,
- non-functional requirements,
- proposed schedule,
- estimated development costs, and
- risks identified and suggested mitigations.

There are many styles and formats for an SOW. Examples and templates can be found at the sites below, or a quick Internet search will provide more targeted examples.

- ***Sample Technology Statements of Work (SOWs)*** @ http://www.gsa.gov/portal/content/133795
- Statement of work @ http://office.microsoft.com/en-us/templates/statement-of-work-TC001184032.aspx

Turn-Over

In some instances the TA team will also be the team to implement. When a different team is used to implement, the TA team should schedule and perform a detailed transfer and turn-over of the project to insure the team implementing understands all aspects the TA team has worked on. The TA team should also be available during the implementation for questions as needed.

Stakeholder Agreement

Contributed by Kevin Lantaff, edp

What is Stakeholder Agreement?

Stakeholder agreement and management of that relationship[1] describes the methods and processes used to identify individuals or groups who may influence the project or who will be impacted by it. It includes the description of each role (contribution and importance) within the context of the project, identifies the requirements for managing the engagement, and specifies communication with and between stakeholders. The end goal is to achieve overall project goals and objectives that lead to project success.

Key Success Factors

The value of stakeholder agreement cannot be overstated. All projects are made up of the sum of its stakeholder parts. If individuals or groups are not properly integrated into the flow of the project at every key step and properly managed to encourage the success of each project phase, the result will be sub-par, if not disastrous. The success of the project hinges on how well the various stakeholders are managed so that the project progresses smoothly, on time, within budget, and most importantly, meets or exceed customer expectations.

Voice of the Customer (VOC)

The success of a project depends on how well the customer has been identified. That includes how well all overt and covert needs are understood and documented, and how well the end product will meet or exceed the customer's need. This is true regardless of whether the organization is in agreement, in word or in practice, with this philosophy. To misstep here and not adequately reflect customer needs with regard to the requirements for the project will result in an unsatisfied customer and damaged customer relationship, as well as time and cost overruns due to scope creep.

The discovery of customer needs and customer advocacy with regard to that need can be the role of several groups within an organization. For transaction document projects managed internally by a corporation internal consultant, business planners or analysts usually take the lead. For communication service providers the responsibility usually lies with customer engagement or implementation managers. In technically-driven organizations the IT or programming department may be responsible. Some organizations employ a Customer Experience Officer for the purpose of adding extra emphasis and focus to customer requirements gathering.

Assistance

Project owners are not always aware of who the stakeholders are, what influence they may have, how important they may be to the success of the project, and how to navigate the political mine field that may exist. For this reason, it is wise to seek assistance from

1 Although project managers have many of the same goals as project owners with regard to overall project success, the duties of a project owner are more managerial in nature. This document focuses more on stakeholder agreement and management from this perspective. For information concerning project manager roles and responsibilities with regard to stakeholder management, see the ***Project Management Institute: A Guide to the Project Management Body of Knowledge.***

those who understand the business structure and how decisions are made, where power centers lie, and the political landscape.

Engagement and Communication

While everyone plays a role, some stakeholders are more critical than others to the success of the project. Putting formal management and communication processes in place that integrate each stakeholder's involvement at the appropriate level is essential. The communication plan should ensure that every stakeholder has an opportunity for their voice to be heard in the appropriate venue, that they get the information they need in a timely manner, and using the method they prefer (the right message to the right person at the right time with the right content).

For instance, executives usually want summary information delivered in a setting and format that makes it easy for them to quickly grasp and absorb the essentials. Project managers and technical leads prefer detail and backup material. Some prefer to have information pushed to them, while others prefer to pull it. Various tools should be employed to ensure all stakeholders are informed. One size does not fit all!

Feedback and Responsiveness

Stakeholders should be given an opportunity to provide feedback at any time. Pro-active efforts should be taken to seek out feedback throughout the project so that issues can be addressed quickly. When issues are discovered with any part of the project, quick action needs to contain the problem and resolve it in the best way possible to ensure project success.

Signoff and Agreement

Although most stakeholders are affected in some way by changes at every project phase, the ones who are impacted the most must have primary signoff authority. For instance, the customer should be responsible for signing off on the final requirements documentation and on the functions delivered at each phase. Service and support teams should have a great deal of influence when it comes to determining the level of product quality and reliability. They should also identify the tools provided in support of service levels. If product test is not measured against product quality as it pertains to customer expectations, then quality may suffer.

Understanding Stakeholders

Navigating the Minefield

A successful project has as much to do with cultural and relational savvy as it does with a comprehensive set of customer requirements and great communication. Understanding how to work effectively within the corporate culture (hierarchy, chain of command, centralized / decentralized power) and how to understand and manage the current political climate, will contribute greatly to the success of the project. Identifying a mentor or coach to assist in this area is crucial.

Learning where the power lies is also a consideration. In general, those with a budget have the power. In financially driven companies the CFO has the power to make or break any project. So understanding how your project will be measured financially is important.

In technology-driven companies, engineers, programmers, and other technologists might have a great deal of influence to the point where what is expedient or familiar to build could take precedence over the voice of the customer. Clear, crisp, thorough, and comprehensive customer requirements are needed to overcome this potential bias.

For companies who have experienced a great deal of litigation, the legal department wields a lot of power. The management team may routinely practice risk avoidance rather than assertively going after market opportunity.

In some companies where mergers and downsizing has occurred or is occurring, HR may hold more power than expected. The point being made is to know where the power lies and prepare for it.

It's Not About You!

Project success not only hinges on your ability to navigate the corporate minefield or understanding the balance of power, but also how well you understand the needs, pain points, and agendas of the stakeholders on your team.

The IT department could be stretched very thin. Their lack of resources may have caused a greater number of defects in the past. Because they are measured on defects they may be unwilling to support your project. They will need a lot of help in the area of value and justification in order to secure more resources. Trying to do more with less in this situation will result in failure.

Upper level management may have close relationships with current vendors and might be resistant to change. Therefore, clear and quantifiable benefits and advantages need to be fully developed and understood.

Manufacturing and Customer Service may have a hard enough time retaining and training high quality personnel with their current workload, that the thought of taking on another project is overwhelming. A well thought out and conservative strategy, timeline, and project plan helps to overcome obstacles.

And depending on how well (or not well) similar projects have rolled out in the past, executives and management in general want to see that your project will rise above and not only learn from past mistakes, but exceed expectations, and meet or come in under budget. Conforming to best practices and understanding lessons learned from the past will help grease the skids.

Stakeholder Interest

One last area of consideration when understanding your stakeholders is their level of interest. Are they supportive, neutral, resistant, detached, or motivated to lead. Understanding this, in conjunction with power, will help in how you manage, communicate, and resolve conflict amongst team members. A stakeholder could have all the power but be neutral in support. Worse, they could be resistant. Or they could be extremely supportive, but lack any power to effect real change or influence others on your behalf.

In general, you want to closely manage those stakeholders who have high levels of power and influence, and just monitor those with low levels of power and influence. Those in between should be kept informed and satisfied.

Transaction Document Stakeholders

The transaction document production workflow has many stakeholders. Determining who they are depends primarily on the purpose of the project. Transaction documents include all correspondence created and produced due to the execution of a business transaction. These include statements, invoices, account opening packages, explanation of benefits, insurance policies (EOC's), informational and collection letters, trade confirmations.

Some business lines are involved in almost every. In addition, each document will be owned by a business area, so they must be involved. Marketing will want a presence with any change to customer correspondence. Some companies have departments or individuals who focus primarily on regulatory compliance or security, and they will need to be represented for regulated and sensitive documents.

Business Owner (planner, analyst, internal consultant): Every document produced is in support of a business function, and every business function has an owner who is responsible for the success of that segment of the business. This person is usually the primary customer who will be ultimately responsible for requirements signoff. The business owner is also responsible for business justification of the project, which drives funding and resource allocation within all departments involved. Examples of business owners are consumer and business credit card market segment managers for the credit card services industry, or HMO and PPO business segment managers for Health Insurance companies.

Executive Owner: An executive is typically assigned to the project to represent senior management. Their role is to support the project at key transition and signoff points between phases or when assistance is needed to resolve conflict. There may be more than one executive assigned from the business and finance areas.

Operations (sometimes known as Document Services):

- **Program and Project Managers:** Companies will assign someone to manage the overall implementation, tasks, and interactions between all of the internal and external stakeholders. Larger companies will split program management and project management responsibilities between two people. Program managers (also known as implementation managers or consultants), acting as an internal consultants and operational subject matter experts, work hand in hand with the business owners (customers) and other key stakeholders to ensure overall project success. Project managers take primary responsibility for task and schedule management, and provide the heavy lifting to ensure all stakeholders are involved at the appropriate levels per the engagement and communication plan.

- **Customer Service and Call Center (sometimes known as account management):** As day to day support for current customers, customer service has a wealth of knowledge concerning how to build and roll out solutions that are easily supported and maximize customer satisfaction. For this reason it is very wise to have customer service involved as early as the business analysis phase. Automated ticketing and job management functions, job status and automated error management, online access to customer documents, and detailed reporting capabilities are just a few of the tools that will enhance customer service.

- **Manufacturing:** Whether manufacturing is performed in-house or is outsourced, a representative must be present, especially during business and technical analysis phases, to ensure document production requirements can be handled by current manufacturing capabilities. If manufacturing cannot support the requirements, then program management and manufacturing must quickly determine a solution for the problem, including cost projections, so that a decision can be made to support the new function or modify the requirements.

Information Technology (IT): As with manufacturing, IT is involved from the beginning to ensure requirements can be supported, and to recommend new technology if requirements cannot be met. Whether outsourcing or producing documents in house, both internal and vendor IT staff must be represented at all stages starting with business analysis.

Marketing: If the new project involves marketing messages (TransPromo) then marketing plays the role of co-owner along with the business owner and is involved during the entire project. But since marketing owns the customer relationship along with sales, and transaction documents touch customers, marketing should always be involved in the lifecycle process of any project that changes the look, feel and content of transaction communications. Customer Experience Officers are usually part of the marketing department.

Chief Compliance and Chief Security Officers: Adherence to security standards is normally controlled by IT and compliance is managed by the program manager or business owner. Some companies establish a special group or person that focuses specifically on these areas due to high risk exposure. Early involvement starting at the business analyse phase is required.

Finance, Accounting, and Procurement: Finance plays a budgetary and cost analysis support role during the initial phases and when scope changes occur due to new requirements or adjustments to cost projections. Procurement is used to help with determining costs going forward for hardware, software and services.

Legal: Projects may involve external vendor contracts and disclosure agreements, as well as intellectual property concerns, which must be settled before vendors can support the project. Document messaging may also require approval from the legal department.

Identifying Stakeholders

Every stakeholder will be involved to some degree, but depending on the type of transaction document project, certain stakeholders play a more active or critical role. In this next section key stakeholders are identified by project type.

Stakeholders by Lifecycle Phase

A project is made up of a series of phases as part of a larger product life-cycle. Each phase constitutes a separate work effort that requires work, rework, approvals and sign off. At any phase along the lifecycle project can also be killed.

Certain stakeholders are more active at various phases or steps along the transaction document life-cycle. The following section covers the customers for each phase, what is accomplished in each phase, and who is involved from a stakeholder perspective.

Requirements Gathering: This is where the customer is identified and requirements are gathered and prioritized. In this phase the why is understood (problem or pain), and the what should be determined (specific requirement or need). The how is addressed in later phases. The voice of the customer is of paramount importance and is combined with market research, focus groups, interviews, surveys, JAD sessions, and compliance documentation to identify requirements germane to the project. Prioritized requirements with a description of high-level benefits are documented for future justification purposes. A document mock-up may be created at this point showing the new product or enhancement being considered.

The line of business owner is usually the primary customer, especially for new projects. Program Managers may lead the effort in this phase working closely with the customer to artfully managing all inputs with respect to felt and hidden needs. Care must be taken not to try and solve the problems in this phase. Doing so will limit the brainstorming activity and overall creative process which is so important when determining customer need.

Business and Technical Analysis: These are two separate phases that overlap significantly with regard to stakeholder involvement. Project scope and success factors, roles and responsibilities of stakeholders, and a complete cost benefits analysis occur at this phase. Due to business or technical limitations discovered in this phase, it is possible that customer requirements may require revision.

Program management works with the business owner to create the categories and framework for the analysis. Finance and procurement, IT, and manufacturing support the team by providing current cost and resource information in order to come to a total cost of ownership for the current state. IT, manufacturing, and customer service assist with determining the technology required (gap) and costs associated with reaching the desired state. Other items such as revenue and profit impact, potential disruptions and risk (economic, security, compliance, etc.), and resource impact are managed by the business owner, program manager, finance representative, and executive management. A complete business case is the result.

Marketing, Sales, and Customer Service, remain advocates for customer experience. Marketing also makes certain brand compliance is maintained and customer messaging is appropriate and correct. Legal creates vendor contracts protecting Intellectual Property and may sign off on proposed messaging. Compliance and Security inject critical requirements. Finance gets involved to ensure costs are fully known and understood.

Stakeholder Adoption: This is the first gate where approval from all stakeholders is sought because the solution building begins. Changes made after this phase can be far more costly than usual unless Agile and Lean management methods are employed. All stakeholders receive a formal presentation along with the final requirements document and cost benefit analysis. A process for change management and final approval is defined and implemented. All legal issues regarding intellectual property, contract terms with vendors and confidentiality are put to rest. Any changes to scope beyond this point require approval from all stakeholders unless other arrangements are made.

Architecture: IT assigns an architecture team and leads the effort, with manufacturing, of identifying and documenting all key processes, workflow steps, and components required from data through delivery, including exception processes and audit points. Project management leads the effort to create the project plan with tasks and timelines for all stakeholders to review and approve. If scope has not changed then key gatekeepers are the business owner, program management (implementation, consulting), IT, and manufacturing, ensuring the solution still meets customer need. If the scope changes then finance, executive management, and the other stakeholders weigh in for approval.

Design: The architectural document and plans are used as a base to create a formal functional / design specification for final review by all stakeholders. This document is usually written by the program manager assisted by IT and the business owner. Other documents such as a document design guide and data dictionary are produced at this phase. All stakeholders, excluding finance, legal, and marketing (if not a marketing project), are the gatekeepers for this phase.

Development: During development there are periodic meetings to review project status and timelines which include internal and external resources and task owners. Several testing phases are implemented to test against customer requirements and provide ongoing feedback. Agile development processes and lean management methods create an iterative or incremental approach to development, testing, and requirements verification. The business owner, program manager, IT, and manufacturing teams are the key gatekeepers. Scope creep may bring other stakeholders back into the approval process. In Agile and Lean environments, gatekeeping responsibility may be given to a smaller group dedicated to ensuring the solution meets the needs of the customer, and bypassing the more formal process outlined in this section.

Test and Quality Assurance: This process often runs in parallel to Development. The testing group is usually part of the current IT and manufacturing teams. In larger organizations this function is performed by a separate group which may serve to produce a higher quality product. All workflow and process inputs and outputs are tested against customer requirements using test cases and production regression tests. Stakeholders are the same as in the Development phase.

Production: Before the solution is released for production, all materials must be purchased against specified stocking levels, resources hired and trained, and a launch date selected. Final approval from all stakeholders is required to exit this phase.

Maintenance: In Lean or continuous improvement environments the maintenance phase is active and fluid. Customer requirements are always being gathered and refined, and the solution benchmarked against these requirements. Mini document lifecycles occur as enhancements are brought forward for consideration and implementation. A small team provides oversight and direction usually involving the business owner, program manager, IT and manufacturing, and others as required.

Stakeholders by Project Type

Technology Refresh or Upgrade: Organizations consider a technology upgrade or refresh to increase productivity, quality, and the ability to better manage the customer relationship. The customer could be any group whose budget is negatively impacted by the current state, or those responsible for generating revenue (quality).

Enhancements to technology require a comprehensive review of current functions supported in order to ensure the new technology or solution will at least meet current needs. If differences or shortcomings exist the stakeholders will need to approve these changes to function and operations. Even version upgrades to existing technology can cause major problems, so care must be taken to test and roll out these new technologies or versions out in a phased approach. The most common examples would be upgrades to, or replacement of, printers, inserters, finishing technology, or major software components (composition, archive, etc.).

Program managers (consultants) play a key role as subject matter experts ensuring existing requirements are met. IT and manufacturing help to make sure existing functions will be supported, the new technology can be supported, and new processes are updated and documented accordingly. Customer Service may be involved if the technology upgrade is significant in scope (as with a move from cut sheet to continuous printing), and may cause customer service disruptions.

Mergers and Acquisitions: This type of project is similar to a technology refresh in that new technology will be merged with old, but on a much larger scale. The customer is at a much higher level and expectations are usually that there will be no business interruptions or outages. For this reason a phased approach to integrating and obsoleting technologies is typically employed.

Business owners will likely be involved as changes to functionality and support may be required (at least in the short-term) and new capabilities offer new opportunities to improve quality, efficiency, productivity and costs (cut sheet to continuous). Program management, IT, and Manufacturing are heavily involved.

Statement Redesign: There are many reasons for statement redesign projects, but most efforts are undertaken to simplify layout and messaging for the purpose of reducing customer calls, improving brand image, or adding marketing messages for cross selling purposes. The marketing department, business owners, and customer service are customers.

An agency may be brought in to perform the redesign. They are usually managed by marketing and the business owner. Customer service plays a very important role to make sure the redesign does not cause an increased number of support calls. IT is involved to perform the actual document redesign within the composition system and make certain document construction and pagination is correct. Manufacturing is involved only if the redesigned statement changes print and fulfillment requirements, such as page and document size or envelope changes.

Regulatory Reporting and Compliance Improvement: The customer in this case is the business owner and upper level management. These groups can delegate some responsibility to a risk and compliance group. Complying with industry regulations not only requires meeting specific requirements, but providing proof that you met the requirement. A detailed audit trail is required for this. For instance, if an Insurance business is required to have the transaction correspondence postmarked (mailed) within seven business days of receipt of data, then not only must the workflow be automated and streamlined to accomplish this, but the business must be able to track each mail piece as it travels through the transaction workflow. They need to be able to act quickly if errors occur, and provide supporting data and reports in the proper format to prove compliance.

Business owners, program managers, and compliance officers form the core team. IT and manufacturing provide technology support to craft the solution along the entire workflow.

Stakeholder Management Process

Stakeholder Responsibility

Every stakeholder has a role to play during the entire project life-cycle.

Team Players

As with any team, the project team is tasked with working together towards a set of common goals. These goals are set by the team and agreed upon early on so that there is no mistake as to what constitutes a successful project.

Every stakeholder is required to meet mutually agreed upon requirements such as meeting attendance, responding to requests, following guidelines and rules set by the team, and communicating clearly, among other things. If at any time a stakeholder is not participating as expected and agreed, actions must be taken to rectify the situation.

Decision Makers and Influencers

Some stakeholders influence decisions and others are tasked with making them for their area of responsibility. Influencers, who are usually subject matter experts, play a critical role in providing valuable input to the decision making process. Decision makers are usually those who have budget authority and whose budgets will be impacted by the project. At any phase all stakeholders can weigh in on issues and pose objections, but decision makers have final say. And, before moving to a new phase of the project life-cycle, key decision makers must sign off.

Final Signoff, Evaluation and Continuous Improvement

Finally, stakeholders are tasked with final approval of the project and decommissioning of existing technologies that have been replaced. They also should participate in post project evaluation activities that seek to improve the quality of future projects by documenting lessons learned and best practices. Participation on an ongoing basis may also be required if continuous improvement is a goal set by the business. Improvements are identified which may kick off a new project life-cycle.

Stakeholder Engagement

A process must be put in place so that stakeholder engagement is managed throughout the document development lifecycle. The purpose is to ensure needs, expectations, and issues are managed in a timely manner. The need to manage stakeholders is greater during the initial phases. Using appropriate commination and negotiation techniques helps to ensure that the effort has a high probability of success.

Stakeholder engagement is managed primarily through status meetings supported by other communication methods. Progress on the development process is discussed using the project schedule, change management document, action item log, and other documents mentioned in the next section.

The key is to present the facts and let the stakeholders decide on the best course if corrective actions are needed. Focusing on facts helps to remove the emotional component of any bad news. The goal is not to find fault, but determine what the issues are, and finding the best approach to correct the issues.

Stakeholder Communication and Documentation

Part of good management is making sure everyone in the boat is rowing in the same direction, so ensuring they have what they need to foster buy in at every step is crucial.

Stakeholder Information

Every stakeholder plays a different role and has different needs within the team. They also have varying levels of importance or contribution along the transaction document production lifecycle. A document called a Stakeholder Register documents all information and communication preferences concerning stakeholders on your team. This is a living document that changes frequently. Due to the sensitive nature of the information, it should be managed by the project owner and shared with program and project managers only.

Information includes:

- **Identification information:** name, title, department, location, contact information, preferred communication methods.
- **Involvement level information:** major role and expectations, power, influence, importance, lifecycle phase involvement, Agile or Lean Management team participation.
- **Classification information:** internal or vendor, and level of support you are likely to receive.

Document of Understanding (DOU)

All stakeholders should approve and agree to support critical success factors for the project as documented in the DOU. This document may be one that is used across multiple projects. Examples of critical success areas in the DOU are:

- **Project charter:** A high level description of the purpose for the project including the problem requiring resolution, high level goals and objectives, and the desired state. This description will be in harmony with the customer requirements section.
- **Customer requirements:** The voice of the customer, reflected in customer requirements, is the foundation upon which the success of the project should be measured. If requirements are not right then nothing is right, and the solution would be very costly to repair afterwards. Most companies spend a third to half the time they should in this area, and a commitment from the team to acknowledge the importance of this step is critical.
- **Justification:** Extracting and quantifying all the value possible is the fuel that drives allocation of budget and staff. Reasonable and solid evidence for project value will secure the resources needed, and all members of the team should contribute and agree. Past project examples of either high or low valuation may help find the right number.
- **Project plan:** Everyone needs to help formulate and agree upon the project lifecycle phases, schedule, and overall scope, including the commitment required from each stakeholder.

- **Resource acquisition and allocation:** All stakeholders should be on board with regard to how the budget is spent and if changes need to be made to project scope if there is a budget shortfall.
- **Ongoing support:** All stakeholders commit to working as a team with regard to cooperation and conflict resolution, appropriate levels of engagement, bi-directional communication, and a commitment to see the project through to success.

Continuous Buy In and Commitment

At all stages of the project lifecycle the owner must continually test the level of engagement and consensus of key stakeholders. Involvement and commitment is measured against roles and responsibilities documented in the Stakeholder Register and Document of Understanding. If stakeholder or organizational behavior requires adjustment, then action plans should be put in place immediately so that the project stays on track. A mentor or coach is helpful here.

Part of continuous buy in is problem escalation or conflict resolution. A clear process needs to be documented so that all stakeholders know how the team will resolve issues. Response to issues should occur as they become known, and resolution sought in a timely manner.

Communication Processes

Communication is the life blood of any project, and is used to make sure all stakeholders are adequately informed so that they can participate to a level that promotes success. Communication is used to develop a relationship connection between stakeholders so that a homogeneous team is formed.

Stakeholders have communication preferences and needs with regard to method, frequency, and content received. Some require detail and others just a summary. Some prefer to pull the information and others prefer it pushed to them. Some wish to have access to all the information pertaining to the project, and others a particular subset. Determining how each stakeholder wishes to communicate and receive information should be logged in the Stakeholder Register, and tools put in place to accomplish this.

- **Communication toolbox:** Communication is difficult under normal circumstance, but is truly a challenge during stressful times. The right balance of online versus face to face communication is different for every organization.
    - **Project sandbox or forums:** Interactive forums are used to share and log key conversations and track the communication string of activities and issue. This tool is usually used by technical staff and those who closely interact with them.
    - **Email:** This is still the preferred method of communication by many. The drawback is that the communication string or history does not reside in a central place for all to see, and history or context is easily lost across collaborators.
    - **Text:** This method is used to quickly respond to requests for information which can greatly speed up the project. Text tools do not provide an adequate audit or history trail for the entire team to see.
    - **Meetings:** If it is true that 90% of communication is inflection, tone, and body language, then there is a 90% chance that your message will

not be heard correctly if you are not in front of the audience. Sometimes organizations iterate electronically for hours or days on a subject or issue when a quick 5 to 10 minute conversation could solve the problem immediately. Face to face meetings are the most effective way to deliver your message to ensure understanding. Video conferencing and phone meetings are the next best thing and are usually more effective than electronic communication.

  - **Online project management collaboration tools:** These tools will allow for collaboration between stakeholders and project personnel by facilitating communication (locally and remotely), assigning tasks, allowing for file sharing, and coordinating schedules and calendars. Team members can share information using electronic comments, forums, text and twitter like function, document repositories, and standard email. Although intended to save time and improve project management effectiveness, some tools can be time consuming and counter intuitive. Lean Management advocates call this phenomenon computer disease, where the use of technology actually hinders communication. Care must be taken to select the right tool that meets the needs of the entire team.

- **Documentation and reports:** A central repository of documents or document sharing solution can be used to share files among team members. Document types are listed below.

  - **Project schedule:** If online tools are used then the schedule is accessed within the tool. Sometimes schedules are exported and posted online as a spread sheet. All tasks, owners, timelines and special comments are placed here, and the schedule is used as a guide to direct and manage team activity.

  - **Change management log:** When scope changes for the project then it should be recorded in this log and subsequently recorded in the customer requirements document. If the change management log is not incorporated into an online tool then the log is usually managed in a spread sheet. Change description, date opened, date closed, owner, stakeholders involved, approval criteria, and status are items tracked.

  - **Action item or issue log:** This log could be managed separately or as part of the change management log under a different category, and is populated with issues or action items discovered during the project management process.

  - **Presentations and meeting notes:** These are stored so that stakeholders and team members can review them at any time and to ensure there is an audit trail tracking action items created during meetings.

  - **Key letters and records:** Critical correspondence between stakeholders and team members may be stored in a document.

  - **Other:** Requirements documents, functional specifications, documents of understanding, data dictionaries, document design guides, legal contracts (internal use only), and document objects and templates.

Legal Concerns

The role of the legal department is to limit or eliminate liability associated with the project. They are involved with any external communication and contract associated with the project. Most legal contracts have to do with outside resources or products used.

- **Confidentiality:** When dealing with outside vendors, including current business partners, a non-disclosure agreement should be in place to protect confidential information and intellectual property. One agreement can be signed by both parties to protect information going both directions.

- **Software licensing:** All software used for the project should be purchased and licensed properly. If a site or corporate-wide license is in place then you must check to see if licenses are available, and if not, ensure they are purchased for the project.

Vendor Contracts

- **Ownership:** If an outside vendor is involved in the development of the solution then ownership of the resulting product must be addressed. Each party could ask for exclusive ownership, or they could mutually own it. Exclusive ownership usually demands a much higher price tag.

- **Acceptance Criteria:** The customer requirements document is usually used as the basis for solution testing and acceptance criteria or a separate document is created for this purpose. Key performance indicators and metrics are also defined and measured before acceptance. These requirements need to be stated in the vendor contract, and is used to determine when the vendor's work is complete.

- **Defects and Service Level Agreement:** Defect levels per a certain number of lines of code, or mean time to fail repair criteria could be listed as requirements. If the vendor will remain involved in a service capacity then service expectations and requirements should be agreed to. This could include problem severity definition, support coverage times and contact methods, and response time commitment.

Architecture

Contributed by William McCalpin m-edp

What is Architecture?

Architecture provides the framework to describe the proposed solution for the project at hand. The definition of the Architecture should occur following the Needs Analysis, Technical Analysis and Business Analysis phases and is the point in the process where the technical and business choices are established.

The technical and business options and then make a choice on which particular technical solution we will design and which budget plan that we will adopt and test our results against.

As William Broddy, m-edp, has noted, "This is where the horse-trading takes place." That is, there must be give-and-take between what is possible and what you can afford to arrive at what the company can settle on as a proposed solution that gives the company the "most bang for the buck."

Architecture is the initial, high-level description of that final proposed solution.

Architecture – Large or Small

The Architecture that describes a proposed solution may be large or small, just as the problems it is intended to solve may be large or small.

Consider the project to make a required regulatory language change to a paragraph in an insurance policy. Because the infrastructure of the solution already exists, the size of the proposed solution to be described is quite small: you change some words in a document object in the composition system's document object library, print it out to see if it displays correctly, then begin the rigorous regression testing to ensure that you have the correct results for each case. As projects go, this requires little thought or decision-making.

Now consider the project that is replacement of a number of home-grown COBOL programs that generate line data with a modern, multi-platform enabled composition product that generates the latest full color print stream with an infinite number of fonts, images and graphics, with add-ons for dynamic composition at run-time, and message management. In real life, such projects totally overturn the document production workflow from stem to stern and can take a number of employees or contractors more than a year to design and implement.

Thus, with Architecture, there is no "one size fits all". Rather, the size of the result of the Architecture phase will be as big as it needs to be.

Financial Side of Architecture

In the Business Analysis phase, we look for the money to underwrite the project by making estimates on how the future system might work and determining ("guessing") how much money the new system might save over a period of time – which money could be applied to the new project as part of its ROI calculation.

However, in the Business phase we can't be sure which technical solution will be adopted so we may have to make a number of calculations based on a variety of technical solutions.

Now in the Architecture phase, based on the technical solution selected, we can develop a budget and review that budget to make sure that it matches the estimates produced in the Business Analysis phase.

In finalizing the project budget we will:

- determine the projected cost,
- determine what we can afford to do (and make sure that one matches the other),
- find funding if necessary for any external expenditures,
- find funding or people for the development project team, and
- secure all approvals to proceed into development.

Technical Side of Architecture

In the Technical side of the Architecture phase we have several tasks to perform in the process of choosing the final proposed technical solution:

- assessment of current technology,
- development of application workflow,
- identification of enhanced or replacement technology, and
- development of the project plan.

These items require a detailed evaluation of processes and practices versus best practices, as well as an impact assessment against other processes or to the resource pool.

It is critical to note that any proposed technical solution as embodied by the Architecture must include an understanding of how the current Architecture functions. How can you determine in advance if the proposed solution will do what you need it to do if you don't know what the current system does, and why?

What Can Go Wrong

Indeed, not understanding what the current system does and why it does it is the source of the most insidious errors in many projects. Because the focus of most eyes in the project is on what is "going to be" it frequently happens that, while the new project correctly delivers on the new features, it inadvertently breaks a one or more important features in the current system causing the new project a serious problem, if not a disaster.

One life insurance company thought that they would replace their 1960s era policy issue system with a more modern system. Their original system was a series of custom programs written in IBM 370 Assembler. The programs were hard to maintain as the original author had left and additional Assembler language programmers were difficult to find. In addition, since it was written for a mainframe environment, there was no built-in GUI for managing the system. All online controls had to be written in a separate CICS system updating a custom database.

The company decided to buy an off-the-shelf client-server based policy issue system. This system came with its own databases and GUIs which more easily fit into the evolving distributed environment that the company preferred. The company spent two years and thousands of hours making the changeover as there was a huge amount of customization that went into the new system.

At the end of the changeover only half of the several hundred life insurance products that had been supported in the old system could be migrated into the new system. Why? Because there were dozens of features in the old, custom-written Assembler-based policy issue system that were not in the new off-the-shelf client server system. Because many of these features had never been catalogued in the old system the lack of these features was not discovered until the formal QA process began.

When it was discovered that a large number of life insurance products could not be generated correctly, a number of previously viable products were dropped and another year or two was added to the overall project as additional code and workarounds were implemented. This caused a huge cost overrun as well as the unknown opportunity lost cost of the previously viable products that were dropped – all because the company didn't fully understand its current architecture before it embarked on the project to completely replace it with a new architecture.

Participants

Just as the size of the Architecture will vary greatly based on the side of the problem being solved, so will the number of participants.

At a minimum you will have:

- the architect / technical consultant,
- financial planner, and
- key stakeholders.

The architect/technical consultant will represent the technical aspects of the proposed solution while the financial planner will represent the business aspects of the proposed solution. Even if there is already an assigned a project manager, that project manager will rely on both technical and business advice in determining and describing the most appropriate solution.

As the size of the project being "solved" by the Architecture phase grows, so does the number of people who become involved. Some key roles include:

- **Document designer:** The person who knows the technical details about either the required changes to the document or about detailed interworkings of the new composition tool being proposed,
- **Database specialist:** For identifying obscure data base fields that will now suddenly acquire a new significance, or for creating a new system or "shadow" data base, or modifying current data bases already in production,
- **Operations analyst:** To review and describe to changed or new manufacturing workflow,

- **Managers of previous and next steps:** For situations when completely new elements are being added to the document composition workflow – such as a new printer or a new composition system,
- **Vendors:** Both hardware and software, as appropriate, as well as any professional services,
- **Financial players,**
- **Document / budget owner:** This would be the lowest level of manager whose budget is responsible for the project,
- **Procurement:** If the project is large enough to require the intervention of a procurement office, and
- **Corporate or divisional finance:** Almost certainly required when enterprise software or any electronic printing/finishing hardware is acquired.

Vendors

In the Architecture phase, vendors can be an invaluable source of information. A vendor who has sold you a substantial amount of hardware or software has a strong vested interest to ensure your happiness with their product. Indeed, the smaller the vendor, the larger the need to keep you happy.

Second, the problem that you are trying to solve has almost certainly been solved by other companies on your industry. That is, for every 1% of the companies that are "bleeding edge", there must be the 99% of the companies that are merely following in their footsteps.

And, if you think about it, when vendors gives you recommendations of fellow user companies to speak to they will give you companies who successfully implemented the new hardware/software. It's one thing during the sales process to want to also hear from customers who had problems with the proposed hardware/software, but once the decision is made, you'll want to know more from the users who made the acquired products work.

With luck, you may be able to get "sanitized" architectural documents through the vendor for your particular project, either because they were created by the vendor's implementation team or were derived (with permission) from the documents of some of their other users, properly cleared of any proprietary information.

In this same way, note that trade associations can help fill in the gap in finding fellow users with similar experiences to the project that you are intending to undertook. While it may happen that there are not any users of the similar hardware or software in your immediate area, trade association meetings can attract regional or even national audiences.

Xplor International is the trade association that represents users and sellers of transaction document production hardware, software, and services. At Xplor events there will be other users of your proposed hardware and software, and there will possibly be open sessions describing how to use the product.

Furthermore, one not commonly discussed advantage of having multiple vendors at a trade exhibition is that the vendors are less likely to be spinning fairy tales to you, since they know that you will be speaking to other vendors just a few feet away…or worse, one vendor is speaking to an audience while on a panel in front of a group of other vendors, who would quickly set the record straight if anything amiss was said.

What is the End Product?

The end product of the Architecture phase is, of course, the Architecture. This is, as noted before, is the high-level description of the single proposed solution that the company has decided to pursue. It may not be the cheapest or the fastest or the technically most sophisticated or even the solution that gets you everything you want, but, at the end of the discussion, you realize that this is the solution that gives you the "mostest" with the resources that you have.

Thus, the end product of Architecture will have:

- budget (i.e., how much will this solution cost to develop, so that future expenses can be measured against it),
- resource plan (what employees and long-term contractors will be need for how long to complete the process),
- new process in terms of workflow,
- clearly and cleanly integrated with current process (yes, don't just describe the proposed process but also describe how it fits in with what you're doing today that is not going to change),
- a description of the expected output,
- specify version / release of software/hardware, and
- generally scope out custom work for either internal or external services.

Who Needs to Approve?

The number of individuals required to approve the results of the Architecture phase will vary considerably, based on the size of the proposed project.

At a minimum, of course, the "owner" of the project must approve the results. This could be the manager of the business unit that owns the document application that was having a small update.

Larger projects will require the approval of each group that is touched by the changes in the project, and in some cases, by groups that are not directly touched by change but who may be affected by upstream changes.

In addition, external groups that are part of the large project planning and execution (such as the project office or accounting office) may also need to approve based on the policies and procedures of the company. Indeed, for company critical projects, the Architecture phase may have to be approved by a senior vice president.

Information Design

Contributed by Robert Linsky

What is Information Design?

For information to have impact, it must be easy to find, simple to use, and instantly understandable.[1] Information design is the defining, planning, and shaping of the contents of a message and the environments in which it is presented, with the intention to satisfy the information needs of the intended recipients.[2]

Why is Information Design Important?

The ultimate goal of any and all communications is to clearly connect the user with the thought. The purpose of information design is to make that communication clear. Utilizing best practices in information design avoids visual confusion, redundancy, and makes content easy to understand. When this happens, effective documents are created.

Best Practices

Information design is informed by a large body of theory and research. It focuses on clear, easy-to-understand language, well-organized content, grouping of like information, use of font weights and sizes to help create a hierarchy of information, graphically-pleasing layout, psychology, and usability. Following best practices in ensures that the result is clear communication. This allows all stakeholders touching a communication to easily find information, understand what they have found, and act on it.

There are many different types of communications and different types of documents; each has a purpose. All documents, whether they are letters, statements, bills, regulatory notices, or forms, fit into the broad category of communications. The difference is the purpose, whether it is imparting information (a letter), requesting information (a form), or communicating status (a statement or bill).

While information design covers ALL communications, the focus of this section is on information and document design narrowed to specific types of paper and electronic communications: letters, notices, forms, bills, and statements.

For any communications to be successful, three things must fall into place. First, a reader can find the information they are looking for easily and quickly. Second, the recipients of the communication can easily comprehend what they have found. Last, the recipients can do something with the information they have found and comprehended.

Users – Psychology

Successful communication requires an understanding of all potential users so that the best practices in information design can be applied appropriately and directed to the stakeholders involved. You may view the stakeholder as *the customer* or *end user*, but most documents affect more than one stakeholder. There are internal as well as

1 Information Design Association (IDA)
2 International Institute for Information Design (IIID)

external users and all of their needs must be considered to create documents that work effectively.

Consider, for example, a bill (utility, telecom, etc.). Before the bill is sent out it must be inspected to make sure the bill generation system is working properly. For data-intensive transaction documents testing the system means ensuring that the data and the calculations are correct.

Then there is the recipient of the bill, who needs to know the amount due, the transactions or reasons for the amount due, when payment is due, and information about late payment penalties. Customer service representatives may need to answer questions from the customer about the bill. The accounts receivable staff that processes the payment relies on information in the bill to apply payment. Each stakeholder is looking for different information so that they can act in an appropriate manner.

On the other hand, consider a hospital pharmacy medication label whose small size creates a challenge to the information designer. Users of the label are all looking for different information: the lab technician identifies use and dosage; the pharmacist compounding the medication; the assistant adhering the labels to the medication; the person delivering the medication to the floor and, don't forget, the nurse administering the medication.

Each document needs to help many people find different information easily and quickly, therefore, all stakeholders must be considered.

Purpose

To be effective a document must have only one main purpose. Too often organizations try to add too much into a document making its purpose unclear. Is it a bill? Is it a marketing piece? What is the recipient expected to do? Many bills look like marketing brochures and some get thrown away inadvertently!

Consider a hypothetical auto lease bill (based upon real examples; see below). It has two sheets of paper, each printed just on one side. Ninety per cent of the second page is blank. The only data on page two is the detail of the payment: principle, interest and monthly loan amount. This is on the second page because there are seven messages on the first page that take up about two-thirds of the page. All are under the heading ***Important Messages***. (More about messages later.)

Generally speaking, only one or two messages are truly *important* and more than that will dilute the value of ALL of the messages. In this case the messages are all valuable, but none are critical or fall under the criteria *important*. They are: view statements online, how to make automatic payments, pay by phone, accident information, mailing payment information, and contacting the company. All valuable, but not worth the extra costs involved.

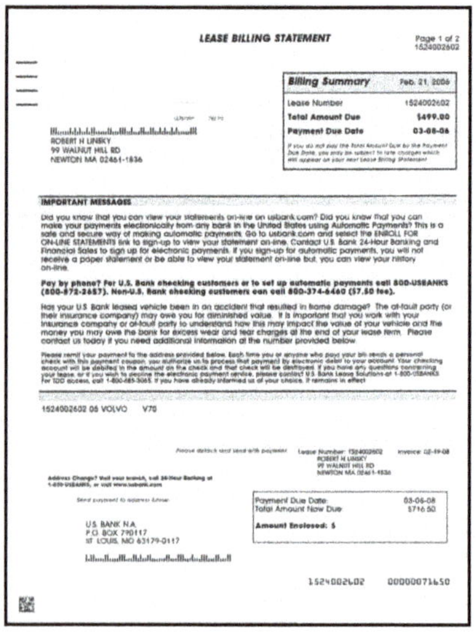
LEASE BILLING STATEMENT

Page 1 of 2
1524002602

Billing Summary	Feb. 21, 2006
Lease Number	1524002602
Total Amount Due	**$499.00**
Payment Due Date	**03-08-06**

If you do not pay the Total Amount Due by the Payment Due Date, you may be subject to late charges which will appear on your next Lease Billing Statement.

ROBERT H LINSKY
99 WALNUT HILL RD
NEWTON MA 02461-1836

IMPORTANT MESSAGES

Did you know that you can view your statements on-line on usbank.com? Did you know that you can make your payments electronically from any bank in the United States using Automatic Payments? This is a safe and secure way of making automatic payments. Go to usbank.com and select the ENROLL FOR ON-LINE STATEMENTS link to sign-up to view your statement on-line. Contact U.S. Bank 24-Hour Banking and Financial Sales to sign up for electronic payments. If you sign-up for automatic payments, you will not receive a paper statement or be able to view your statement on-line but, you can view your history on-line.

Pay by phone? For U.S. Bank checking customers or to set up automatic payments call 800-USBANKS (800-872-2657). Non-U.S. Bank checking customers can call 800-374-6460 ($7.50 fee).

Has your U.S. Bank leased vehicle been in an accident that resulted in frame damage? The at-fault party (or their insurance company) may owe you for diminished value. It is important that you work with your insurance company or at-fault party to understand how this may impact the value of your vehicle and the money you may owe the bank for excess wear and tear charges at the end of your lease term. Please contact us today if you need additional information at the number provided below.

Please remit your payment to the address provided below. Each time you or anyone who pays your bill sends a personal check with this payment coupon, you authorize us to process that payment by electronic debit to your account. Your checking account will be debited in the amount on the check and that check will be destroyed. If you have any questions concerning your lease, or if you wish to decline the electronic payment service, please contact U.S. Bank Lease Solutions at 1-800-USBANKS. For TDD access, call 1-800-685-5065. If you have already informed us of your choice, it remains in effect.

1524002602 05 VOLVO V70

Lease Number: 1524002602 Invoice: 02-19-08
ROBERT H LINSKY
99 WALNUT HILL RD
NEWTON MA 02461-1836

Address Change? Visit your branch, call 24-Hour Banking at 1-800-USBANKS, or visit www.usbank.com

U.S. BANK N.A.
P.O. BOX 790117
ST. LOUIS, MO 63179-0117

Payment Due Date:	03-06-08
Total Amount Now Due	$716.50
Amount Enclosed: $	

1524002602 00000071650

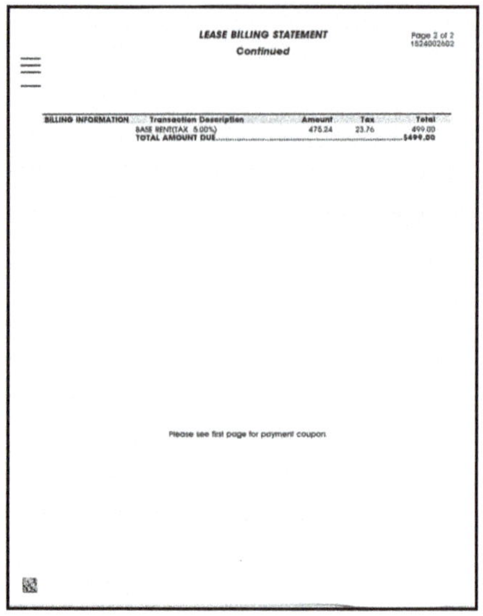
LEASE BILLING STATEMENT
Continued

Page 2 of 2
1524002602

BILLING INFORMATION	Transaction Description	Amount	Tax	Total
	BASE RENT(TAX 5.00%)	475.24	23.76	499.00
	TOTAL AMOUNT DUE			$499.00

Please see first page for payment coupon.

Figure 1- Sample Generic Leasing Statement.

Messages

Messaging is very important and can be a useful tool in marketing and maintaining customer loyalty, but it must be done appropriately. There is a fine line between success and failure in messaging. If customers receive monthly statements with the same message repeated two or three times in a row, not only will they stop reading the message, but future messages will be passed over, too. In the example above a better decision would have been to spread the messages over multiple months. This would achieve four things: there would be a fresh message every time, each message would stand out by itself and become more important, there would be cost savings due to fewer sheets of paper, and there will be a better customer experience.

Document Design

Most people are familiar with pure graphic design – making the document or marketing communications look good. But many of these *pretty* documents fail because they do not communicate or, in the case of the electronic document, cannot be rendered. How often has a company *designed* a form, bill, or statement only to find that the development tools or the printers cannot produce it because it does not fit within the parameters of the equipment?

Document design incorporates information design, plain language, psychology, and usability testing in addition to pure graphic design, and – especially for transaction documents – includes the understanding of the environment in which the document lives. Too often organizations look at documents as nothing more than a cost of doing business, either because they are regulatory or they are fill a need, like a bill, which helps to fund the organization. Although this is true, good document design, using the best practices of information design, goes beyond look and feel to functionality, cost reduction, and increased customer loyalty. Therefore, involving information designers at the start of any customer-facing document project is essential. Relegating information design until the end of the design project, with the only intent to make the document pretty, is useless.

Most transaction documents include variable data. Think of a utility bill, a credit card statement, or a retirement or bank statement. They all look alike (from a graphic design and branding standpoint), but each is unique because of the data in the document.

Variable Data in Transaction Documents Methodology

The information designer brings a lot to the table when designing variable data transaction documents. In the beginning of the process, when content is on the table, a knowledgeable designer can help visualize the ways in which it can be presented. This not only will optimize the design and development, but ensure the document works for all stakeholders. Unlike a graphic designer, the information designer will understand the development and production requirements, and restrictions. They will also understand that transaction documents need to allow for differences in content within a particular field and allow for that variation.

Color

Color grabs your attention and quickly conveys important information; imagine traffic lights without red, green, and yellow! Judicious use of color can be used to emphasize important information, but overuse will defeat that purpose.

Color should be used carefully, not only to make an attractive document, but also to serve a higher purpose: clarifying information. When using color, consider consistency so that similar information is emphasized throughout the document. Take the example of the traffic light. Everyone knows that the red is on top and green on the bottom, but think about what would happen if they were sometimes reversed. That would make for a lot of confusion, traffic jams, and accidents. The same is true of documents. If a bar chart showing one type of information has three different colors and then a pie chart uses the same colors for different types of information, the stakeholder will be confused thinking that there is a connection between the two visuals. See Figure 2 as an example.

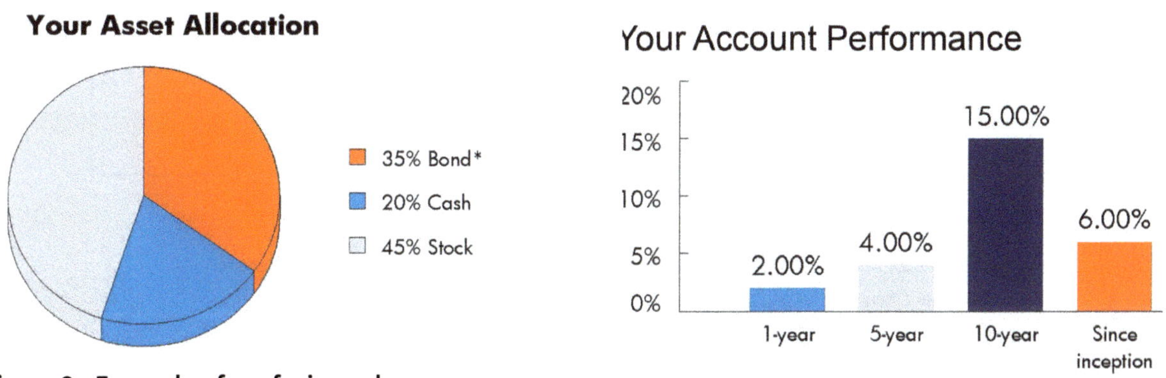

Figure 2 - Example of confusing color.

Proper use of color can emphasize important information, increase the reader's attention, leave a lasting impression, and increase comprehension.

Fonts

Fonts play a large role in transaction documents. It is important to choose a font that is readable in different sizes as well as having a variety of weights (bold, italic, condensed) since most transaction documents will be data-intense and font weights will be used to create hierarchy and visual separation for readability.

Readability is improved significantly when the number of font families used is limited to a maximum of two. Except in rare cases most information designers avoid using type smaller than 9 point. This will depend on the communications and the type of content. Most content should be in the 10 to 12 point range. For certain regulated documents the smallest allowable font size may be mandated. For instance, in a financial statement with many sections and numerous columns within a section, 10 point might be the largest that will fit.

For sections that contain a substantial amount of numerical information, a sans serif font is most readable. By using different weights and different point sizes for different types of information, a clear hierarchy will be created. But, consistency across the document is important, just as it is when using color. For fonts, using the same weight and size for similar information is very important.

The before and after example below clearly shows the value of using different font sizes and weights to visually separate information. The before example uses one font weight and one size (8 point) while the after (taking up less space) utilizes bold that emphasizes different information from the regular weight and various point sizes (larger – 10 and 12 point).

ANTICOAGULATION (BLOOD THINNER OR BLOOD CLOT PREVENTION) MEDICATION AND TREATMENT INFORMATION	☒ You are being sent home on medicine to prevent blood clots (also known as anticoagulation): **Indication:** INDICATION FOR ANTICOAGULANT MEDICATION DOCUMENT. **Expected duration:** EXPECTED DURATION YOU'LL TAKE THE ANTICOAGULANT. ☒ You are being discharged on the blood thinner Warfarin (also known as Coumadin). ☒ Overlap with Warfarin/Coumadin and Enoxaparin (also known as Lovenox) is planned. ☒ You are also being discharged on these oral blood thinners: Aggrenox(DipyridamoleASA) Arixtra (Fondaparinux) **Target INR Range (Blood level that tells us how well your medication is working)**:350-400 **Your next blood test should be drawn on :** 07/23/08 **Contact and location information for blood test:** YOU SHOULD MONITOR THIS REGULARLY WITH YOUR MD **Name of physician or clinic that will follow you:** MR. ROBERT TOURNIQUET - 617-555-0120

Anticoagulation (Blood Clot Prevention) Medication and Treatment Information

You are being sent home on medicine to prevent blood clots (also known as anticoagulation).

You are being discharged on the blood thinner Warfarin *(also know as Coumadin)*.
Overlap with Warfarin/Coumadin and Enoxaparin *(also known as Lovenox)* is planned.

Indication:
Indication for Anticoagulant medication document

Expected duration:
Expected Duration you'll take the anticoagulant

You are also being discharged on these oral blood thinners:
Aggrenox *(DipyridamoleASA)*
Arixtra *(Fondaparinux)*

Target INR Range:
350-400 *(Blood level that tells us how well your medication is working)*

Your next blood test should be drawn on:
07/23/08

Contact and location information for blood test:
You should monitor this regularly with your MD

Name of physican or clinic that will follow you:
Mr. Robert Tourniquet - 617-555-1020

Figure 3 - Before and after example of good use of fonts and white space.

White Space and White Space Management

An important concept to understand, and the least understood, is white space and white space management. White space refers to the areas in a document without text or images, just the background. Organizations tend to want to *fill* the page. In reality, white space allows the eye to travel from section to section, from concept/idea to concept/idea, easily and quickly, making the document easier to understand and information easier to find. Look at the example above. Notice that the *before* example is almost completely filled up, but in the *after* example the white space visually separates the content making it easier for the eye to follow.

Barcodes and Other Codes

Transaction documents many contain many different types of marks and codes. All of these have to be incorporated into the information designer's list of items to be considered. There are barcodes, like *3 of 9* and *2D*, *IMB* for postal regulations, and others for inserting, and page count. Other than the postal codes, there are usually options for placement and it is the job of the information designer to work with the equipment requirements so that the codes serve their function without interrupting the flow of the content and readability.

INFORMATION DESIGN PROCESS

Requirements Gathering

Gathering requirements for an information design project is a vital first step. The gathering of requirements becomes a valuable internal communications tool because it should succinctly capture the document strategy, objectives, and priorities.

This phase addresses specific business, marketing, and production requirements and includes items such as product features, existing document data, corporate branding and image guidelines, record-keeping systems, and the target document delivery environment. The analysis of this information provides the basis for the document strategy and the parameters under which the document will live. For the communication to be produced in the transaction world this is a vital step. If done at the beginning before any design it will help save time, rework, and expense.

Document Strategy

The strategy addresses high-level design and content objectives for the document and establishes the *stake in the ground* against which detail content and design options will be evaluated and measured. The strategy objectives will be used to inform the content and design concept.

Usability Testing

There are many types of usability testing: attitude surveys, preference tests, focus groups, expert opinion, and diagnostic testing to name a few (some of these have other names). Focus groups are the most popular, but for information design, diagnostic testing has been proven to be most successful. For financial and other personal documents specifically, diagnostic testing using one-on-one protocols are recommended. This method is used because the results are more indicative of how individuals view them. In this type of testing, a moderator is employed to interview a participant with questions and scenarios. In this way, testing will reveal how well the participant can find information, understand information, and navigate through a multipage document. This testing is done with these three questions in mind:

1. How accurately does the document communicate information?
2. How well does it guide the users?
3. How well does it help people in using the document to solve a problem or answer a question?

If the budget allows, testing should be done with current state samples as well as the proposed design. This is to make sure all issues and problems are identified before any redesign is started. A second testing is done after the redesign. This is to verify that all user requirements have been captured and all user problems and pain points have been identified. A document must reflect how people think about a task, how they find information, and what they do with that information. Therefore, testing should be done with actual users, both internal and external.

Testing will determine how accurately the document communicates information, how well it guides the user, how users actually use the document to verify information or answer a question, and will find out what final improvements are needed. The document is successful if it naturally fits into work routines like any well-designed tool.

Content

Before any graphic (look and feel) design is started, the content must be defined. Developing the content first will save rework and keep the redesign within the project timeline. In this phase, existing content is reviewed to make sure it is appropriate, not redundant, and whether it will be valuable to the stakeholders.

New content is considered through the lenses of stakeholder's needs and/or regulatory requirement. The key consideration is to identify where this new content is coming from. Content is also reviewed and validated to ensure that terminology is consistent throughout the document and that plain language is used wherever possible. Here, also, is where messaging is included, not necessarily the actual wording, but instead, identifying the messaging type (marketing, regulatory, disclosure, commentary, etc.), how many places it would be used, and where throughout the document it would be placed.

Specifications

In any transaction document project, creating functional specifications, measurement specifications, and font specifications will solidify the design, form the basis for the development, and ensure that the final outcome will stay true to the efforts put in by the whole team. The functional specifications include business rules, widow and orphan rules and other guidelines that help to define how the document will act in various scenarios. The measurement specifications will help the development team by providing exact placement of the different elements of the document, including address blocks. The font specification details all of the font types, weights, and points sizes used throughout the document.

Validation

Since transaction documents involve many people, departments, and processes, it is important that each document is validated throughout the project. First, review content with business units, stakeholders, legal and compliance among others. For the design review, marketing should also be added to the list. And, the development team becomes very important to verify that the design content is available, the design can be implemented in the desired tool, and it will print correctly on the desired output platform.

Because transaction documents are generally complex documents, with lots of information and sometimes many, many pages, validation along the way will keep the project on time and avoid costly rework.

The conclusion is clear: information design and clear communications will create a good customer experience and ensure the successful outcome of any transaction document project.

What is Good Document Design? – An Example

In many respects, a well-designed document represents a compromise between all of the document stakeholders.

- Marketing will be concerned with branding, imaging, and messaging elements, while graphic designers will be concerned with aesthetic elements such as layout and imaging.
- Customer service and accounts receivable will be concerned that critical data points like amount due, due date, account balance, and transactions are accurate, easily identifiable, and readable.
- Compliance departments will be concerned that legal notices, regulated text, and disclaimers are accurate for each jurisdiction to which a document might be sent and rendered in the mandated font sizes.
- Developers will be concerned that all elements in a document are placed correctly, variable data fields are filled correctly, business rules are applied properly, and that all finishing marks and barcodes are placed properly on each page.
- Recipients will be concerned that they receive the correct information at the correct time, and that they can find the information that is important to them.

For many, describing the attributes that represent a well-designed or poorly-designed document is not an easy task; however, most individuals have no problems identifying well-designed or poorly designed documents when they see them. The following bank example uses a variety of techniques to create clear communications that results in good document design. In addition to color, fonts, white space management, and plain language, creating successful document designs involve understanding all the stakeholders and the purpose of the document.

Ultimately, good document design works because it allows the stakeholders to easily locate information, understand the information found, and then act on that information.

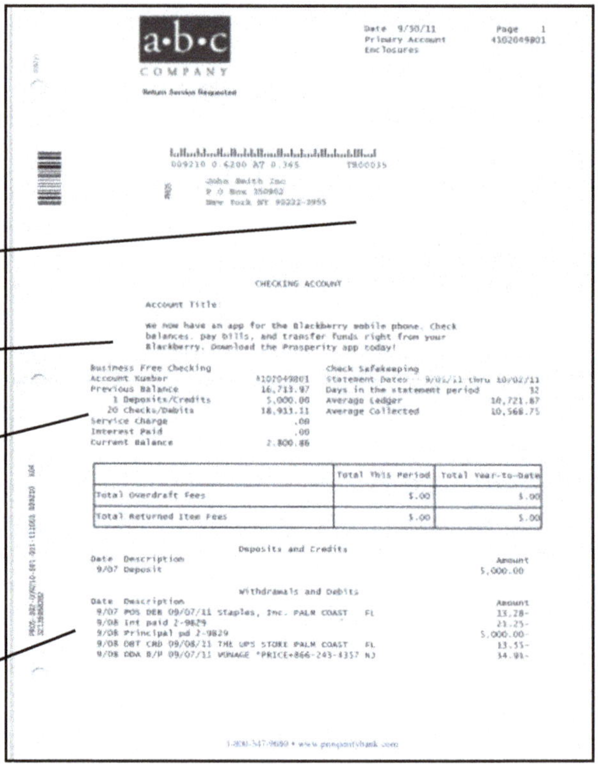

Date 9/30/11 Page 1
Primary Account 4102049801
Enclosures

a·b·c COMPANY

CHECKING ACCOUNT

Account Title

We now have an app for the Blackberry mobile phone. Check balances, pay bills, and transfer funds right from your Blackberry. Download the Prosperity app today!

Business Free Checking		Check Safekeeping	
Account Number	4102049801	Statement Dates	9/01/11 thru 10/02/11
Previous Balance	16,713.97	Days in the statement period	32
1 Deposits/Credits	5,000.00	Average Ledger	10,721.87
20 Checks/Debits	18,913.11	Average Collected	10,568.75
Service Charge	.00		
Interest Paid	.00		
Current Balance	2,800.86		

	Total This Period	Total Year-to-Date
Total Overdraft Fees	$.00	$.00
Total Returned Item Fees	$.00	$.00

Deposits and Credits

Date	Description	Amount
9/07	Deposit	5,000.00

Withdrawals and Debits

Date	Description	Amount
9/07	POS DEB 09/07/11 Staples, Inc. PALM COAST FL	13.28-
9/08	Int paid 2-9829	21.25-
9/08	Principal pd 2-9829	5,000.00-
9/08	DBT CRD 09/08/11 THE UPS STORE PALM COAST FL	13.55-
9/08	DDA B/P 09/07/11 VONAGE *PRICE+866-243-4357 NJ	34.91-

Poor use of white space.

Message area is buried and loses its impact.

All content looks the same – same font, weight & point size It is impossible to distinguish different types of information easily and quickly.

Separate sections – can be combined to better clarify content.

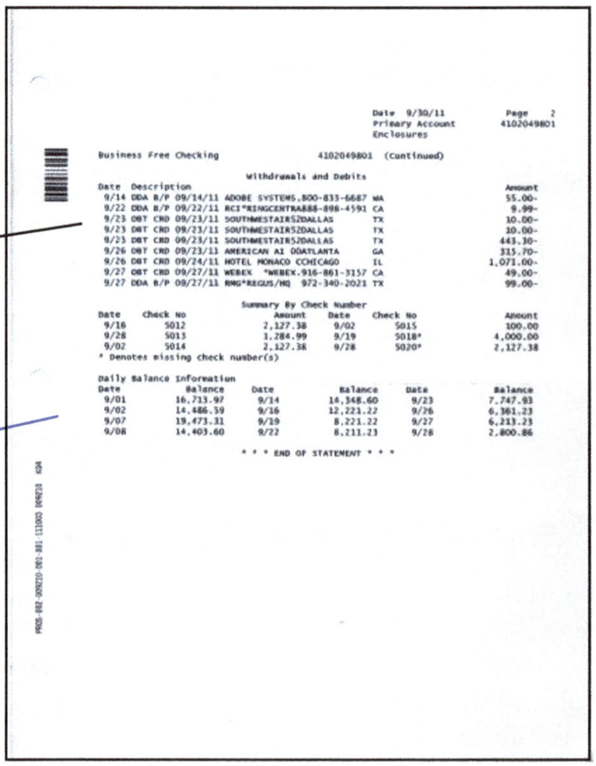

Date 9/30/11 Page 2
Primary Account 4102049801
Enclosures

Business Free Checking 4102049801 (Continued)

Withdrawals and Debits

Date	Description	Amount
9/14	DDA B/P 09/14/11 ADOBE SYSTEMS,800-833-6687 WA	55.00-
9/22	DDA B/P 09/22/11 RCI*RINGCENTRA888-898-4591 CA	9.99-
9/23	DBT CRD 09/23/11 SOUTHWESTAIR52DALLAS TX	10.00-
9/23	DBT CRD 09/23/11 SOUTHWESTAIR52DALLAS TX	10.00-
9/23	DBT CRD 09/23/11 SOUTHWESTAIR52DALLAS TX	443.30-
9/26	DBT CRD 09/23/11 AMERICAN AI 00ATLANTA GA	315.70-
9/26	DBT CRD 09/24/11 HOTEL MONACO CCHICAGO IL	1,071.00-
9/27	DBT CRD 09/27/11 WEBEX *WEBEX.916-861-3157 CA	49.00-
9/27	DDA B/P 09/27/11 RMG*REGUS/HQ 972-340-2021 TX	99.00-

Summary By Check Number

Date	Check No	Amount	Date	Check No	Amount
9/16	5012	2,127.38	9/02	5015	100.00
9/28	5013	1,284.99	9/19	5018*	4,000.00
9/02	5014	2,127.38	9/28	5020*	2,127.38

* Denotes missing check number(s)

Daily Balance Information

Date	Balance	Date	Balance	Date	Balance
9/01	16,713.97	9/14	14,348.60	9/23	7,747.93
9/02	14,486.59	9/16	12,221.22	9/26	6,361.23
9/07	19,473.31	9/19	8,221.22	9/27	6,213.23
9/08	14,403.60	9/22	8,211.23	9/28	2,800.86

* * * END OF STATEMENT * * *

Account identity & section headers are difficult to see.

Two-column format takes up more space & adds to page count.

No color used.

Figure 4 - Before example of information design make over. These images are not crisp due to being scanned.

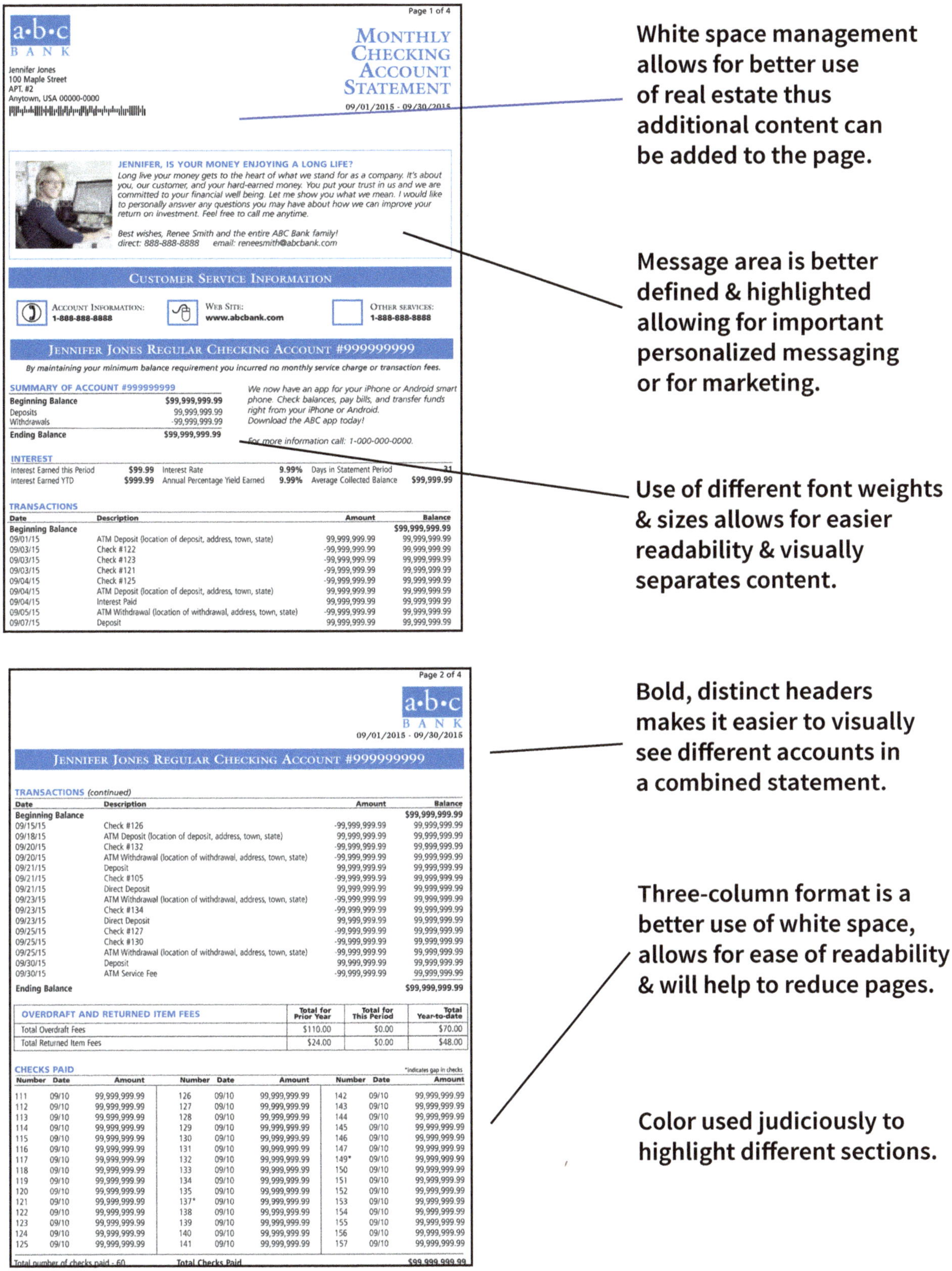

Figure 5 - After example of information design make over.

(Project) Design

Contributed by Denise Miano edp; Cheryl Kay; Wendy Macmillan m-edp

What is Project Design?

The Design phase of project is an important phase. It finalizes, at a detailed level, the requirements, specifications, solutions and approach for the next phase of Development. Depending on the type and size of the organization and the methodology used to manage and resource projects, detailed specifications may already have been produced and signed off during the Analysis phase. When this is the case they are received as input along with other related project documentation and reviewed during this phase.

Many companies produce detailed specifications during this phase. This is the reality of their process, resourcing, skillset and their relationship with their client and/or project sponsor. This phase is written from the perspective of identifying what must be finalized as the output from the Design phase.

If detailed specifications are produced during this phase it may be necessary to circle back to earlier phases of the lifecycle, such as architecture, business analysis and technical analysis, since fleshing out the detailed specifications here may identify a new business and/or technical requirement that impacts the cost, timeline, and resourcing. Skilled project management is essential in identifying these possible impacts and managing them.

Establishing the Project Team

Some organizations assign a Project Manager early in the lifecycle to manage the activities related to moving a project from an idea through costing and pricing, business and technical analysis, solution definition, and architectural review especially for large and complex projects. Others assign a project manager when the project is approved and funded and moves into the Design phase.

It is imperative in this phase that there is an identified and engaged multi-functional team. The team should have representation from each department / business group that is affected by the project. The team should include decision makers and workers; with the decision makers acting as a steering committee. There should be an executive sponsor to ensure that the project is resourced and allowed to proceed without roadblocks.

The following roles should be involved in the Design phase of the project:

- Project Manager
- Project Sponsor
- Business (Systems) Analyst (either full-time to develop detailed business requirements or part-time to provide clarity)
- Data Specialist
- Workflow Specialist

- Technical Leads with expertise in all technical solutions being proposed.
- Architect
- Application Developers
- Technical Services (Hardware, software, communications , connectivity)
- Operations
- QA Analyst
- Customer Stakeholders including Decision Makers
- Procurement

The skills needed by the Project Team to produce System Requirements are dramatically different from those required to write and then translate the business requirements into a technical design. Business requirements define the what; technical and system requirements define the how.

Procurement may need to be involved if the project requires the purchase, installation and training related to new/additional/upgraded hardware and/or software. The installation of hardware or software may comprise the entire project, or mean a prerequisite larger project. Examples might be the installation of a new printer or inserter to handle growth in print volumes or the installation of a new composition system (including resource hiring or training) to deliver a new statement project.

Not all initiatives are large statement engineering projects and they may not be a Project at all in the formal sense of the word. Many organizations deliver far more change requests than they do large scale projects and the Design stage may be fairly brief, identifying the current state accurately and then designing the future state and how it will be achieved.

Other projects in our industry might include adding a new channel, such as messaging or e-presentment to an existing solution; de-commissioning an application; upgrading software or hardware; rebranding output to reflect a merger or acquisition; moving from one software or hardware platform to another; or implementing workflow automation. The nature of the project will drive the exact nature of the approach and documentation in the design phase. This section should be reviewed with that in mind.

It is common for the team size to expand as you progress through this phase to be sure that the range of skills is adequate for the task. There can be an advantage to keeping some of the original team intact while progressing to this phase to retain continuity in the business knowledge and functional expertise from earlier phases.

At this point in the project any training and support needs must be identified, as well as all equipment and communications needs.

Design Steps

The following steps must be completed during this phase to ensure that the planned solution meets the defined requirements at a detailed level:

- All documentation prepared at earlier phases is reviewed to ensure the business rules and requirements and other specifications are defined in sufficient detail. If not, activities are planned and executed to obtain the details in this phase.

- Project documentation is either reviewed or created to ensure clear understanding by all team members of the project purpose, scope, assumptions, timelines, volumes, SLA, deliverables and budget. This documentation should also include the following information, gathered during the previous phases. If it does not, this must be captured and documented in this phase and perhaps be captured as part of the Business Requirements.
  - **Overview:** This section reflects the intention and describes the project at a high level.
  - **Problem Summary:** This section summarizes the basic problem that the customer faces. It includes specifics regarding the customer's current business scenarios. It also describes the challenges and areas of particular risk. It may also point out areas that are working well so there is minimal disruption to those areas if possible. Those that are working well may also help to pave the way for the eventual solutions.
  - **Objectives:** This section contains a clear statement concerning the primary goals and objectives of the project. This is a very important section as it will be used to guide the success of the project. Here is where a company should include input from all stakeholders. If possible, the objectives should be measurable. This will help with reporting and benchmarking the solution. Key items to consider are:
    - SLA requirements,
    - Transaction volumes,
    - Transmission frequency,
    - Recovery Time Objective (RTO) and Recovery Point Objective (RPO) for contingency planning, and
    - Security requirements.
- Major Required Features: This section starts out with a high-level, prioritized statement of the features that the project requires. From there, the details of the features are reviewed including information regarding how to implement the features to meet the needs of the customer. The output from the Architecture phase is reviewed to understand how existing and/or new enterprise technologies or solutions can be leveraged for the current design in areas such as composition, data normalization, print and insertion, messaging, archive, epresentment and other channels. Gaps between installed and required capabilities are identified and a plan put in place to address them.
- The requirement for new hardware or software to support the solution is identified and a project created for procurement, installation, training and support.
- An in depth analysis of all the data structures, forms and business rules. At this stage, similarities, differences, level of complexity and relevance are identified. The goal is to use this information to start to frame the implementation approach and confirm the architecture for the new solution. . The recommendations from this phase will lay out the structure, strengths and functionalities of the software and platform.
- Once all specifications are produced in sufficient detail to either confirm the planned approach or identify gaps, the Project Manager produces detailed task level project plans and a detailed resource plan and reviews all project assumptions and scope to ensure they are still valid.

- Changes are identified and the impact on project timelines, resourcing, cost and budgets are identified. At this point a project cost estimate from previous phases may be refined from a range (e.g. $100 to $150K +/- 20%) to a fixed or more closely defined number for stakeholder approval.

Specifications

The final specifications or outputs from this stage, whether simply reviewed from previous phases, updated or built from scratch, are passed on to the Development phase for the solution to be coded, configured, installed, and/or unit-tested based on the requirements. These requirements are also used as the basis for building plans for testing in the QA and Test phase through traceable testing. They must be written clearly and concisely and the QA and Test Analyst should be consulted about an effective format for writing and archiving of documentation.

If the project is a change request to an existing application, current state documentation should be marked up to demonstrate the future state in appropriate sections to ensure that the exact implications of the change are captured and the documentation remains current for future change management.

Business Requirements should always be reviewed, approved and signed off by the client or sponsor. These are the requirements that will drive development and all the phases that follow and allow Project Managers to manage changes that impact timelines, resources and costs.

Outputs may include the following:

- **Data Dictionary:** The analyses of all the data structures involved in the creation of the documents. This will include evaluating the detail down to the field level, the field name, its source, field characteristics, type, and content. The work product from this analysis is the critical predecessor for documenting the business rules and creating a comprehensive mapping of the business rules to the data. Based on this analysis, it may be possible to break the implementation into phases based on features like data structures, document types and frequencies and internal priorities.
- **Business Requirements:** For all pieces of the solution. These may include desired features and operations, screen layouts, business rules process diagrams, pseudo code, file naming conventions and other documentation, Business rules that drive the content of a complex document such as a statement can be very lengthy and detailed and require considerable time and facilitation with business users, who may require time to investigate and make a decision. This area should not be underestimated in planning.
- **Document Design Guide:** This is often produced by a design or marketing department or a third party design firm that designs the look and feel, information flow and functionality of a new or re-engineered document. The guide will include information such as logos, pictures, font types and other document objects, style usage, colors, page and table break guidelines, wrapping, column formatting and other functionality. It will also include a mockup of the document. If this guide does not exist, this information must be included in the Business Requirements.
- **Style Guide:** If the company sponsoring the project has a Style Guide that defines the use of their logo, branding, fonts, colors and so on, this should be collected and provided as input to the Development phase.

- **Configuration Guides:** For existing solutions or products in the current architecture, such as archive or epresentment, configuration guides are produced that define how built- in functionality features are set for this particular project. These are produced in consultation with the business owner and require good knowledge of the solution and good communication skills.
- **Functional Requirements:** for features such as sorting and bundling, postal sorting, data normalization, special handling, manual processing and so on
- **Operations Requirements:** for bar codes document integrity, marketing inserts, envelopes, paper, audit and reconciliation, SLA and file splits to accommodate postage tiers, convenience volume splits, work flow etc. More information about these features can be found in other sections of this document, but at the design stage the specific requirements to allow Operations to handle the application in their environment accurately and efficiently must be clearly defined at this stage.
- **Technical and Support Requirements:** For such items as networking, data receipt, passwords, security protocols, and error reporting.
- **Hardware and Software Installation Requirements:** For new or upgraded hardware and software projects. This will include items such as floor plans, HVAC and other environmental controls, technology platform specifics, software versions, training, testing, and stress testing.
- **Technical Architecture Definition:** The purpose of the Technical Architecture is to describe in more detail the high level architectural workflow identified during the Architecture phase and to ensure that the Technical Solution fits into the Enterprise Architecture of the organization and meets the requirements of the project. The technical solution includes information such as terms of the hardware platform, programming development languages, supporting toolsets including third party tools that are to be used in the creation and maintenance of the new solution. It should also document interfaces to databases and other input sources, document objects, archives, preference management and messaging tools, billing, audit and reconciliation and various outputs such as print, archive, epresentment and accessibility formats and also identify various environments for Development, User Acceptance Testing (UAT) and production.

Development

Contributed by Neil Merchant m-edp

What is Development?

The Development or Build phase in the systems development lifecycle is where the project deliverables start life. In the preceding steps the project objectives have been outlined, the business case developed and the detailed analysis conducted. Design has laid down the overall structure of the solution and produced a detailed set of application, software and/ or product specifications. Now the construction starts.

In a hardware installation project it is getting the equipment on the ground, laying it out, installing power, air-conditioning and waste extraction, and getting it all ready for integration testing. In application development, it is the file/database creation, coding, composition or manipulation tool configuration, JCL writing and unit testing. A commercial software development project uses a similar process for application development

Hardware Installation

Preparatory Work: Any necessary planned upgrades to air conditioning, power supplies, waste extraction and other infrastructure should be completed before equipment installation.

Equipment Delivery and Assembly: Any old equipment should be relocated, or decommissioned and removed, once any remaining workload has been successfully migrated to other hardware. It may be necessary for old and new hardware to co-exist for a period of time, which should be allowed for when planning the use of floor space.

New equipment should be delivered ready for installation. Once onsite it can be moved to its target position as defined in the Design phase (make sure floor loading has been checked), assembled and connected to waste extraction, power and servers as appropriate.

Service clearances should be checked and floor markings (for walkways, clear areas etc.) painted.

Commissioning and Setup: Once installed, the equipment will be subjected to a range of commissioning tests by the vendor. Inserters will be configured for planned jobs and those configurations saved. Printers will be put through a variety of tests which will vary based on the print technology.

Training: If the equipment is new to the customer or has new operator features initial training should be delivered by the vendor. Out of this training and unit testing, new or revised operational procedures should be prepared.

Unit Testing: Once commissioning and training are complete, the vendors will hand the equipment over to the customer, and unit testing can begin. This typically involves test runs using copies of existing work or standard test packs. The goals are to make the operator familiar, to compare quality and performance, and to identify initial problems.

Any new stocks, inserts, and envelopes should also be tested at this stage, to ensure: compatibility to specifications; reliable processing through the new equipment without wrecks and jams; and ink or toner performance.

Temperatures around new equipment should be checked to identify any hot spots and allow remedial action.

At this point, hardware installation is complete, and the project can progress to the Test phase.

Application Development

The development team's job is to take the architecture and design specifications that result from the identification of business requirements analysis and build an executable solution. These specifications may define the software to be used, programming languages, database tools, hosting and other constraints as well as the process structure and flow (i.e. the steps in the job and the structure of individual programs.)

Reuse of common modules and exposed services may also be included in the specifications. If not, care should be taken in the Build phase to identify such opportunities.

Architecture and Design may have identified the need for new commercial software (composition or manipulation tools). If so, the initial installation and testing of these may better be considered as a separate project: selection and acquisition may take several months, followed by installation, training and a pilot application. Developing a mission-critical application at first use brings significant additional risk to the project schedule.

Most applications consist of three functional layers – data, business logic and presentation. This is certainly true in transaction document applications, even though business logic and presentation are generally achieved in the same (composition) step. These three layers are discussed below.

Data: Data in this context consists of two major elements: document objects, used repeatedly across many documents, and variable data, unique to each document.

Document Objects: Document objects will generally be prepared or sourced for reuse during this phase. These objects may include the set of fonts specified in the document design guide, company or product logos, blocks of text defining product features, legal terms and conditions, customer servicing and other details, and barcodes and other document integrity marks. These may need to be produced in versions for print, web and mobile use, and all will need to be unit-tested.

Variable Data: Given the volumes typically involved, and the potential diversity of sources (e.g. account database, customer database, CRM systems, marketing/targeting systems) variable data extraction is an important element to get right. It is typically handled in the first step in the overall process, generally known as Extract, Transform and Load (ETL). Poor choice of tools, and inefficient use of appropriate tools, can lead to excessive processing times that can jeopardize meeting service level agreements.

Normalization and pre-formatting of currency amounts, telephone numbers, addresses and other data elements is good practice at this stage, and also provides the opportunity for data validation (for example confirming that all necessary data elements are present, and checking for reasonableness in currency amounts). Intelligent integration and structuring of extracted data from the various sources into the input for the composition or manipulation step can again provide performance efficiency.

Many composition toolsets offer ETL capabilities that include adapters for popular database systems (Oracle, DB2 etc.). Depending on the complexity of the need these may be fit for purpose. In more complex, demanding or unusual circumstances it may be better to develop the ETL functionality in-house.

Business logic such as sorting/splitting for service level, mailing and print operational reasons may also be applied at the ETL stage, as can that for defining additional content based on the account holders behavior as manifested by the transactions incurred. This content can either be included later by means of a metadata trigger added to the account's data records, or pulled in at this stage.

Careful attention to developing data integrity checks during the application's Build phase is also key to a successful, high quality deliverable.

Business Logic

Business logic, the set of business rules, is generally applied at the composition or manipulation step, and implemented using the tools provided by the composition or manipulation product set. The majority of these product sets are functionally rich with graphic front ends that greatly simplify the development process. They are far more complex than products such as the MS Office suite, and training in their use is essential. If the tool is new to the development team, professional services assistance from the vendor for the first project is advisable. Learning hands-on how to use the product from an expert in accordance with its design philosophy rather than applying your own preconceptions r misconceptions will pay dividends.

For composition, a data dictionary is generally built first, identifying the contents of each input data record type field-by-field, establishing relationships between them, and assigning attributes.

The document can now be built using the designs and style guide that will have have been developed much earlier in the project, usually with a GUI-based tool. The elements of the document – address box, product and account information areas, and tabulated transaction areas are built up along with functions such as pagination rules. Elements from the data dictionary can be positioned appropriately on the document design, and processing rules applied using the composition tools functionality.

The specifications may also call for other outputs to be emitted from the composition process, such as integrity or control files for managing the downstream print, insert, mail, and email processes and to help track the processing of each document throughout its journey. Again, current composition toolsets have the capability to create these.

In the case of a print-stream manipulation project, the requirements are generally simpler. These are a few examples:

- reversing the order of documents in a file,
- address validation and correction; adding or changing barcodes, and
- use of white space to include marketing content.

As with composition, the tools now available are generally sophisticated and functionally rich, and the required manipulations can generally be achieved without resort to writing traditional code. Service providers, for example, often use print-stream manipulation tools to re/format the bar codes and other document integrity from "print ready" files received from clients to their standard.

Presentation

Where does the presentation layer begin?

It begins somewhere in the composition step.

Most composition tools separate the application of business logic and the creation of presentation streams. They generate an intermediate internal file format containing presentation intent, which is then processed into specific presentation formats for different languages and media: print-streams (e.g. PDF, AFPDS, PCL); email notification (HTML, text); and for serving to web and mobile presentation (typically XML or PDF) where the data is stored and then served up on demand (i.e. it goes back into the data layer). One of the joys of composition tools is that the actual generation of a presentation stream comes "in the box", which eliminates the need for programming work. It only requires specification via an option or parameter setting, and possibly the purchase of one or more additional emitters.

JCL/Control Scripts

To run unit tests of the developers' deliverables, control scripts are required to run the application. In a mainframe environment these are usually JCL (Job Control Language) scripts, in UNIX environments shell scripts (*.sh, *.ksh, *.bsh) are usually used, and on a Windows platform these may be Batch Files (*.bat). For the Test phase more extensive scripts are required to run the entire suite, including allocating output files, clean-up and error handling. These scripts may be written during Build or it may be an early activity in the Test phase.

QA Activities

Typical quality assurance activities in the Build phase include re-use checks, code and configuration reviews, document layout reviews, and integrity checks (document counts).

Re-use Checks

Whether developing in-house ETL code or using composition/manipulation tools, there may be reusable modules, data dictionaries, design templates or entire applications available for reuse. In the case of ETL, for example, the formatting of data elements (date, currency, etc.) will likely be consistent with other documents, and that formatting code should be reusable. With good prior design of data and templates, there may be very little coding work needed to implement a new products statement into an existing family.

Code and Configuration Reviews

Code walkthroughs with suitably skilled colleagues not otherwise involved in the project can highlight misinterpretations of specifications, pick up coding errors and introduce better ways of achieving the desired functionality.

Walkthroughs with composition toolsets are more difficult, as much of the work is interactive at the toolset GUI and not amenable to review: here, the vendor's professional services team may be of help.

Document Reviews

Although the range of output available from the (unit) testing conducted during Development will be limited, reviews against the document design checking for print placement, formatting, and pagination can highlight overt errors and confirm the interpretation of the specification.

Unit Testing

The role of unit testing is to demonstrate that the unit of code (program, composition rule-set, etc.) behaves according to the specification, and is ready to go forward to subsequent, more extensive, testing procedures. What follows is a basic example to demonstrate the purpose; unit testing will be designed to match the specific circumstances of ever project and piece of code.

Valid files: Testing with valid input files will demonstrate correct behaviour and the outputting of files for subsequent printing or validation.

Empty files: Testing with one or more empty input files will confirm correct behaviour – i.e. detection of the empty file and an orderly termination with a suitable error message.

Data out of range: Testing with input files that contain data out of range in one or more ways (e.g. address line length, currency amount in excess of the value specified, date invalid or unreasonable, phone number with too few/many digits, number of transactions in excess of specified limit)

Missing data: Testing with input files that, for example, have no street address (for a document destined to be mailed), no customer name, or a missing final balance amount, will demonstrate that these conditions too have been checked and correctly catered for.

All required output types: Test output should be generated for every desired language and target medium, and rendered if possible at this stage.

Integrity checks (document counts): Integrity checks throughout the document journey are essential in the production of variable data documents. At every stage, documents should be counted, and the counts passed on to the next stage for comparison with subsequent counts. Although it's not generally good practice to conduct financial calculations within document systems, hash totaling transaction amounts to check against the final balance supplied from the systems of record is another way of catching errors.

Control file checks: If control files are included in the specification, their content should be reviewed in detail. At this stage it may not be possible to test them directly with the downstream steps in the process, but they should be validated as far as possible.

Phase Deliverables

The deliverables from the Build phase are reports documenting and confirming that the phase activities and tasks have been completed successfully. Depending on the nature of the project, these may include:

- all hardware installed successfully and subjected to initial tests,
- sample base stock, inserts and envelopes received and tested,
- application code written and unit tested,
- software, such as composition or manipulation tools, configured and unit tested,
- all output from unit test reviewed and checked,
- list of QA activities conducted and of any errors or problems detected and fixed, and
- checklist of phase gating criteria and compliance.

Critical Communications Recovery

Contributed by Christine Durfee

What is Critical Communications Recovery?

Critical communications and your company's print and mail operations are vital to your organization's survival. Invoices, statements, policies, payroll – if you cannot print and mail these documents, your company can suffer fines, loss of cash flow, or compliance issues. Having a recovery plan in place for these documents is essential. With the variations and intricacies in today's print environment, recovery efforts must meet the specific requirements of your company to be successful. Having the right plans in place, which often means working with a recovery partner, are the milestones to a successful recovery.

This high level overview reviews the steps to take to develop, test, and maintain a critical document recovery plan.

Business Impact Analysis: Defining Requirements

A Business Impact Analysis (BIA) is a tool used by continuity and recovery professionals to assess a company's disaster preparedness and resulting financial risks relative to a major business disruption. If your company has completed a BIA, the risks of losing your critical document production facilities should have been detailed and you already have an understanding of the recovery scope needed to ensure that your company's production needs can be met. If you are starting from the ground up and need to provide upper management with evidence of organizational impact should your critical document processing facilities suffer a business disruption, the following are some high level questions and scenarios to review.

Organizational Impact

Consider what types of business disruptions may affect your ability to produce your critical printed documents. Natural disasters, power outages, pandemic, and work stoppages are overall concerns. In a print and insert production environment, technical issues must also be noted as potential disaster scenarios. Interruptions to printer or inserter hardware or software due to technical difficulties can affect your company's ability to meet your Service Level Agreements (SLA's), may last longer than expected, or cause such stress on your capabilities that your recovery partner may need to assist.

In a total loss scenario, replacement of production equipment must include realistic timelines. The days of replacing a printer in one to two weeks or an inserter in four to six weeks are long gone. Specialized equipment requires additional manufacturing time and set-up once delivered. Planning for a lengthy outage to your production facilities will impact your recovery scope. With a long-term disaster both high-level critical documents and communications that are lower on the priority list must be included in the planning.

Critical Documents

What documents are crucial to the continued success of your company? How long could your company manage its day-to-day operations without the flow of your critical communications? How will your company:

- maintain cash flow,
- keep customers informed,
- maintain regulatory compliance, and
- avoid service level penalties and fines.

All customer-facing communications must be reviewed for inclusion in your recovery plan based on these considerations.

Business Units Print/Mail Expectations

Determining what documents should be skipped or included in your recovery plan is a monumental task. All company departments that utilize or are affected by these documents should be included in the recovery scope discussion.

Once the critical documents are identified, priority and service level requirements must be detailed. Remember to include start-up time at the recovery partner site. Once a disaster is declared, data, forms, envelopes, and other materials must be transferred to the print and mail recovery facilities. These delays, although part of the normal recovery process, will affect your standard SLA's and must be taken into account. In addition, consider set-up time at the recovery facility. Although their printers and inserters may equal the production speeds of your own, they cannot go from zero to one hundred in one day. Documenting these start-up obstacles will provide realistic expectations for all parties.

Internal Recovery Considerations

A successful critical communications recovery plan requires inclusion of all business units working with, or affected by, print and mail operations in recovery planning discussions. They must all have knowledge of the plan details and objectives. Units that process or generate printed documents may include:

- IT – Main Processing Site
- Mid-Range, Server, Satellite Sites
- Business Units
- Support Groups
- Remote Offices
- Organization Partners
- Processing Centers

Multiple Print and Mail Operations

If your company operates multiple print and mail production sites and utilizes these facilities as their recovery plan, consider these questions to determine if you can meet the criteria for actual recoverability.

- Do they print and/or insert similar jobs?
- Are the facilities mirror images of each other (data formats utilized; print hardware and software; insert hardware, software, barcode methodology, and finishing requirement)?
- Do they routinely transmit jobs between sites for processing?
- Are the facilities running at or near capacity?
- Are they staffed for influx of additional production volume?
- Can all jobs be produced at the alternate sites?

Answers to these questions will highlight if your multiple locations can truly act as recovery sites for each other. As with all recovery plans, testing of data transfer, print, and insert capabilities should be done on an annual or semi-annual basis to assess the continuity plan and reveal any hidden gaps.

Outsourcing

If your company outsources any or all of your critical document production processes, it is important to understand that compliance and any service level agreements are ultimately your responsibility. The recovery of your critical documents should be managed and maintained by internal controls and processes rather than those of your outsource vendor.

Outsource vendors with multiple facilities may provide a recovery solution to you that includes the use of their secondary facilities. If so, be sure to use the checklist above regarding multiple locations to confirm that, in the event of a disaster, your production requirements can be met by the outsource vendor's secondary location(s).

If your outsource vendor operates only one facility and does not have a formal back-up or business resumption plan, it is imperative that you work together to find a recovery solution that will meet your print and mail requirements. In this instance, bringing in a third-party print and mail recovery partner may be your best solution.

Print and Mail Recovery Requirements

After completing your Business Impact Analysis, you should have an understanding of the critical documents that will need to be replicated in the event of a business disruption. Discuss the following details with your recovery partner so that they have an understanding of your print and insert needs. Some specifications are more crucial for implementation rather than at start-up, but the more information you can provide to your vendor at the beginning will allow for a more thorough understanding of your environment.

Print Equipment and Volumes

Provide a detailed list of all printer types (and quantity), models, and specifications that reside in your production environment.

Provide a spreadsheet of print volumes required for recovery. This may not represent all print produced in your facility, but should include critical applications identified during the impact analysis phase. Volumes should be provided as total images (8 ½" x 11" (or A4) images or impressions) and should be grouped into the following sub-categories:

- Cut Sheet (indicate if simplex, duplex, or both)
- Continuous (indicate if simplex, duplex, or both)
- Highlight Color Cut Sheet (indicate if simplex, duplex, or both)
- Highlight Color Continuous (indicate if simplex, duplex, or both)
- Full Color Cut Sheet (indicate if simplex, duplex, or both)
- Full Color Continuous (indicate if simplex, duplex, or both)
- Any MICR requirements

If there are print jobs that use legal size or non-standard size pages, note them.

Print Data Transfer

There are several options for transferring critical print data to your recovery partner. Your options will depend on your circuit availability (FTP, VPN, T1, etc.) and your print and/or host systems. The following is a list of options that your print and mail recovery environment/vendor should be able to support. If your data and print environment requires something non-standard, discuss options with your recovery planning team. Provide your recovery partner with the types of host systems that drive your print process. There may be more than one type. Some common options:

- IBM Mainframe
    - JES2/JES3
    - PSF /Infoprint Manager / Process Director
    - Prisma
    - Custom Statement Formatter (CSF)
    - XPAF
    - VPS
    - JQP
- Operating System Level – ZOS V_.R_
- User Modifications / System Exits
    - JES2/JES3 exits
    - PSF exits

    - VPS/JQP exits
    - Other exits____
- Unix/Linux
    - SUN
    - HP
    - Hitachi
    - Generic OS Level
    - IBM I-series OS Level
- Windows/Other OS Level

Print Data

Provide the types of data streams used in your print environment:

- Line Data
- LCDS
- Metacode
- AFP / IPDS
- PDF
- Postscript
- PCL
- Unformatted Other_____

Print Software and Processing Reconstruction

The recovery plan requires details of any print management tools involved in your print process. The following identifies some of the common tools, but consult with your print operations team to ensure what tools are in use, the release levels of the tools in use, and what additional scripts might me in place. All of these must be documented and available in the recovery environment.

Crawford Technology Production Manager	LRS VPS / VPX	Pitney Bowes JESConnect	Solimar SOLitrack
Canon Océ Prisma	Ricoh InfoPrint Manager	Pitney Bowes FlexServer	Solimar SPDE
GMAC Software	Ricoh TotalFlow Producer	Pitney Bowes VIP Server	Rochester Software QDirect
BARR Host Comunication Suite	Ricoh InfoPrint ProcessDirector	Xerox XPAF	MPITech
BARR Enterprise Print Server	Mirus Remote Sync	Xerox FreeFlow	

Are any of the following applications used to generate the print data stream :

GMC Software PrintNet	HP Exstream Dialogue	EMC Document Sciences Compuset / Autograph	Xerox VIPP
GMC Software Inspire	ISIS Papyrus DocExec	EMC Document Sciences xPression	Xerox XICS
Thunderhead	SCRIPT-based command languages (DCF / BookMaster / Waterloo Script)	COBOL or other programmed formatting	Xerox XMPie
Custom Statement Formatter (CSF)	DocuMerge	Napersoft CCM Author	Bitstream Pageflex
Pitney Bowes Doc1	Objectif Lune / Printsoft PRES / New Leaf	Objectif Lune PlanetPress	EZ-Letter
Inventive Designers Scriptura Engage	Tranformations Uluro	Doxee	

Are post-composition re-engineering products in use?

Crawford Technologies Product Suite	Solimar Product Suite	Sefas Product Suite	Actuate Product Suite
TargetStream	Rochester Software Product Suite	Pitney Bowes Product Suite	Canon Océ Prisma
Ricoh Infoprint Product Suite	Compart Product Suite		

Remember to describe any special processing tools for managing data to the printer that may not be detailed above. This includes network to mainframe protocols, legacy resource files like electronic overlays, and other special requirements.

Insert Equipment and Volumes

Provide a detailed list of all inserter types (and quantity), models, and specifications that reside in your production environment.

Provide a spreadsheet of insert volumes required for recovery. This may not represent all insert/mail packages produced in your facility, but should include critical applications identified during the impact analysis phase. Volumes should be provided by envelope size and should be grouped into the following sub-categories:

- #10 (indicate average page count per envelope)
- 6" x 9" (indicate average page count per envelope)
- 9" x 12"or other flat size (indicate average page count per envelope)
- Pressure Seal Requirements
- Hand Inserting or Special Packaging Requirements

Any special processing completed on your inserters should also be identified, including two- or three-way matching, selective inserts, stapling, booklets, and post-insertion imprinting.

Insert Technical Details

Along with providing the types of insert equipment used in your facility, your recovery partner will need to know some further specifics in order to replicate your production environment. These details can include the following:

Feeding	Face-up / Face-down
Edge Leading	Top / Bottom
Collation	A to Z Z to A
Address Location	Top / Bottom
Print Orientation	North/South East/West
Average Page Accumulation	Page Count
Maximum Page Accumulation	Page Count
Fold Type (Letter)	Z Half Roll fold Other______
Barcode Configuration	2D 3of9 2of5 OMR(4th 6th 8th) Other Logic_________
Envelope Size	Identify size
Envelope Type	Window Closed Self
Imprinting Requirements	Identify

A copy of your barcode methodology is needed by your recovery partner to assist in understanding how your applications are assembled on the inserter (number of pages per envelope, selective inserts, match requirements, etc.) and any document tracking requirements.

Postal Requirements

Postage is a significant segment of your print and mail costs. Ideally any postal savings applied during normal operations should be duplicated in the event of a business disruption. Your recovery partner (internal or external) should have the ability to sort mail online, or have access to a pre-sort vendor that can apply the optimal postal discounts. Discuss your postal strategy with your recovery partner and confirm that they can match your efforts.

Develop a plan for moving postage money from your local accounts to the recovery partner. If metering is required, an allocation of funds via check or electronic funds transfer (EFT) to a meter account will be needed. If you are using indicia (permit) for your mail, your recovery partner must check with their local post office to ensure your permit will be allowed to mail out of the recovery location at time of disaster.

Quality, Reprints, Waste

Your recovery partner should have their own standard quality control procedures, but in the event of a disaster declaration your operational procedures should be maintained as closely as possible. It is important to share your company's quality requirements with your partner, even internal partners, so they may follow your procedures and ensure the integrity of your printed and inserted documents.

A process for handling reprints should also be discussed. Since most recovery sites will work with print image files (instead of raw data), they may be unable to reproduce the document at the recovery facility. Verify during drills that all file types can be processed as needed.

Any waste as a result of print or insert technical issues must be reported for inclusion in the next day's print volume. Exact procedures for reporting waste details will vary by organization and may include reporting a sequence number, name and address, account number, or some other identifying information on the damaged document.

As with any document containing personal information, waste material must be handled as securely as possible. Your recovery partner must demonstrate the ability to contain all waste in secured receptacles and offer document shredding services, either on-site or offsite, with reporting documentation. The security of the personal information on the printed documents is ultimately your responsibility and therefore every measure taken in your own production environment should be followed at the recovery site.

Security and Compliance Requirements

Security and compliance is a major part of your company's day-to-day operations and must also be part of business continuity plans. Your recovery partner must maintain the same security and compliance requirements as your own print and mail production facilities.

Compliance requirements will vary based on your core business and geographical location. For example, HIPAA regulations are required for the healthcare industry and PCI certification is required for the financial industry. Your company's compliance manager will have the security details you must follow and which will be required of your recovery partner. Most companies will require some type of security and compliance questionnaire to be completed by an outside vendor upon selection, with yearly updates needed. The questionnaire may also be required of internal recovery partners.

Vendor Criteria

If you are choosing a vendor recovery partner, choosing the right vendor is critical to assuring the continuity of your print and mail operations. A recovery partner should have facilities dedicated to handling print and mail production requirements, or at the very least dedicated equipment to handle a potential disaster. Look for the following as you evaluate your potential partner:

General:

- Recovery as their core business, industry knowledge, and history,
- Dedicated recovery facility and/or dedicated equipment capabilities,
- Multiple locations or secondary back-up facility,
- Business continuity plan in the event of a disruption to the recovery site,
- Disaster declaration experience (how many, how long),
- Standard test procedures and processes along with the ability to follow your specific test requirements, and
- Security qualifications similar or higher than your own: ISO, PCI, GLB, HIPAA, etc.

Facilities

- Multiple communication transfer options for delivery of data from your facility to print and mail recovery facilities,
- Printer hardware, software, and volume capabilities to duplicate your production requirements,
- Insert hardware (with any special functionality, i.e. barcode methodology, matching, booklet making, stapling, etc.) and volume capabilities to duplicate your production requirements,
- Document tracking procedures equal or similar to your environment,
- Ability to meet your postal requirements (pre-sort, permit mail, manifest, etc.) with postal facilities on-site or near facility capable of handling additional mail volume,
- Document shredding for secure handling of all production waste,
- Daily tracking reports and procedures for notifying you of project status,
- Standard quality control procedures, and the ability to implement any specific requirements you may need,
- Dedicated staff: Customer service, IT, Network, Print, Insert, Postal, Quality, and
- 24/7 operations.

Recovery Plan Implementation

The implementation of your print and mail recovery plan can begin as soon as your recovery partner has an understanding of your print, insert, and mail specifications.

Your implementation team should include members from the following departments: DR/Business Continuity, IT, Network, Print Operations, Insert Operations, and Postal. If you use outside vendors for data recovery they should be part of initial implementation to ensure successful data transfer.

Your recovery partner may request an onsite review of your production facilities. This visit can be an important step in understanding your specific print and insert production requirements, giving your vendor the tools to better duplicate the recovery effort in their own facility.

The following is a general list of action items you should expect to complete for a successful implementation of your recovery plan. Some items may not apply to your environment, and there may be other requirements not detailed below.

	Critical Communications Recovery Implementation Checklist		
	Action Item	Responsibility	Completion Date
	Proof of Concept Test		
❑	Print and Insert Grids to Print/Mail Recovery Vendor	Customer	
❑	Pre-Test Discussion to determine test scope, expectations	Print/Mail Recovery Vendor	
❑	Review data, print and mail specifications with Customer	Print/Mail Recovery Vendor	
❑	Barcode logic and sample MRDF files to Print/ Mail Recovery Vendor for test set up	Customer	
❑	Pre-Test Samples to Print/Mail Recovery Vendor	Customer	
❑	Proof of Concept Test at Print/Mail Recovery Vendor	Print/Mail Recovery Vendor	
❑	Proof of Concept Test Report to Customer	Print/Mail Recovery Vendor	
	Account Implementation - General		
❑	Review non-tested application specifics	Customer and Print/ Mail Recovery Vendor	
❑	Gather samples and detailed specifications of all non-tested applications	Customer and Print/ Mail Recovery Vendor	
❑	Gather additional barcode logic or MRDF files (if needed)	Customer and Print/ Mail Recovery Vendor	
❑	Quality Control Procedures to Print/Mail Recovery Vendor	Customer	
❑	Additional security procedures required (if any) to Print/Mail Recovery Vendor	Customer	

	Critical Communications Recovery Implementation Checklist		
	Action Item	**Responsibility**	**Completion Date**
❑	Declaration Authorization Information to Print/ Mail Recovery Vendor	Customer	
❑	Review postal procedures for disaster declaration	Customer and Print/ Mail Recovery Vendor	
	Data		
	Description of Data Transmission method requested		
❑	Configure and test method of transmission	Customer and Print/ Mail Recovery Vendor	
❑	Finalize automation of file receipt decryption and spooling notification, if needed	Print/Mail Recovery Vendor	
❑	Confirm resource requirements, set up monthly transfer, if requested	Customer and Print/ Mail Recovery Vendor	
	Print		
	Description of Print Requirements		
❑	Determine initial control document for print priority (at time of disaster)	Customer and Print/ Mail Recovery Vendor	prior to next test date
	To understand what files are being sent to recovery facility.	Print/Mail Recovery Vendor	
❑	Maintain print specifications on print grid, samples, etc.	Customer and Print/ Mail Recovery Vendor	
	Insert		
	Description of Insert Requirements		
❑	Maintain insert specification on insert grid, samples, barcode methodology, postal requirements, etc.	Customer and Print/ Mail Recovery Vendor	
	Account Maintenance		
❑	Change Management Updates	Print/Mail Recovery Vendor to Customer	
	Print/Mail Recovery Vendor will send Change Management memo for updates on changes to volume, equipment, applications, etc.		(more frequently upon client request)
❑	Inventory	Print/Mail Recovery Vendor	
	Annual Testing		
❑	Determine annual testing schedule (quarterly, semi-annual, etc)	Customer	upon receipt of contract
❑	Send test date requests to Print/Mail Recovery Vendor	Customer	upon receipt of contract

Testing

Testing your print and mail recovery plans will assure that your recovery partner can support your requirements if and when you need them. The more you test, the more comprehensive your tests are – the smoother your recovery will be.

A typical print and mail recovery exercise should include the following production processes: transfer of data to the print and mail recovery facility, printing, and inserting of identified critical applications. Standard test procedures should include a variety of print and insert applications to insure you can match all of your requirements.

A Proof of Concept test should be performed prior to recovery plan acceptance. Yearly, semi-annual, or quarterly testing should be completed to maintain recoverability and test any changes to your environment that take place throughout the year. It may be wise to schedule ad hoc tests when new applications are added to production.

It is important to combine your data recovery test with your critical document testing to insure seamless transition of data from your normal operations to the recovery location. Your data recovery tests should include backup and recovery of all print systems and applications so that you can confirm all recovery time objectives. If your print and mail recovery plan calls for transmission of data to the print recovery site within eight hours, but the data recovery plan priorities or timeframes do not match this requirement, the success of your print and mail operational recovery is at risk. Work with your IT and Data departments to detail your critical applications and timeframes so that recovery scope is understood by everyone involved with the process. Remember that there may be regulatory requirements that guide recovery times.

Disaster Declaration

Your print and mail recovery partner should maintain formal disaster declaration procedures, with contacts and instructions for movement of materials, postage, and other requirements. Your internal recovery plans should include who at your organization is authorized to formally declare a disaster and who will manage the critical document recovery efforts. Appropriate staff members at the company and the recovery partner should be identified and all contact information should be exchanged.

Materials Management

Movement of required materials for print and insert production should be detailed in your recovery plans. It is suggested that you manage stock levels at your print sites so that in the event of a disaster you can request immediate shipment of materials to the print and mail recovery site. Most recovery sites have warehouse availability, but due to paper sensitivity it is not suggested to keep materials onsite for long periods of time. However, keeping a small amount of material is suggested for set-up and testing purposes. Discuss your material requirements (paper, forms, inserts, envelopes) with your recovery partner to determine your best options for quick delivery.

Plan Maintenance and Change Management

As technology evolves, changes to your print, insert, and mail processes are inevitable. Modifications and upgrades to data or communications environment, printer hardware and software, insert technology, and postal requirements will impact your business continuity plans. It is vital to report any changes in your environment to your recovery partner as soon as possible as they may require alterations to your service. Revisions should be communicated to your recovery partner as they happen, but it may help to set up quarterly or semi-annual review meetings with internal departments and your recovery partner to discuss any changes to your environment.

Summary

Print and mail production environments have evolved into technical wonderlands that can no longer be easily replicated at the time of a disaster. A recovery plan is essential. Understanding the requirements needed to produce your critical documents in the event of a business disruption, partnering for print and mail recovery that can match your specific needs, and testing these recovery efforts can be a cumbersome task, but a necessary one. As compliance and service level requirements become more intense, recovery efforts must match the ever-increasing demands of your internal environments. The right plan + the right vendor = recovery success.

If you would like to learn more about critical document recovery, the following resources may help you find a consultant or vendor that can assist: http://www.drj.com, http://www.continuityinsights.com.

Test and QA

Contributed by Neil Merchant m-edp, Tim Ciceran edp

What is Testing?

Quality Assurance (QA) defines activities conducted throughout a project to ensure the accuracy and completeness of the project execution. Examples of those activities include ensuring that:

- All stakeholders are identified and engaged,
- Requirements are completely collected and accurately documented,
- Hardware, software or application design is appropriate, robust and compliant with regulations and internal standards,
- Standard project management processes are followed, and
- Disaster recovery requirements are included and addressed.

Testing is the subset of QA that relates to proving that the new or modified product (hardware, software, application) works as designed, is safe, is reliable, and is fit for purpose. Some organizations have a separate QA department or team. When this is the case that team may conduct their own tests on the product in a phase known as QA Testing or Acceptance Testing.

Even the testing should be subject to QA, however. QA for testing should verify that all processes and procedures have been followed. It should also ensure that all records and reports have been completed and that all requirements have been tested and proven.

Purpose of Testing

Testing is required to ensure that all business processes and product interactions continue to function as required and that new ones work as designed. Changes in composition software, workflow software, transforms, print hardware, vision systems (e.g. bar-code readers), and even paper can subtly alter the environment. At a minimum the following items should be part of an overall testing protocol:

- business, legal and compliance requirements are accurately expressed in the product,
- the product operates reliably,
- special, exceptional and error conditions are managed,
- applications run within the time available,
- the product cohabits satisfactorily with the existing architecture portfolio,
- there is enough IT capacity (processor, memory, disk, bandwidth) to handle the workload,
- the print room can manage and process the printed output,
- non-print delivery channels maintain all appropriate interfaces and formatting,

- the operations group can run and manage the product reliably,
- integrity checks work,
- paper types and forms work reliably,
- all finishing works as planned, and
- mailstream insertion is not impacted.

Managing the Testing Process

Most organisations will have a standard development methodology that includes work objects (e.g. requirements definition templates, report templates), processes and procedures to lead and guide the project team through all the activities in the testing phase. If these work objects are not in place, this is a good time to assemble them and make them part of the testing protocol. A goal of the testing process is to ensure consistent planning and execution.

Test Strategy

A test strategy defines the high-level plan for testing that is appropriate to the product and project. Defining a test strategy embraces:

Where	what locations, test environments
What	up- and down-stream limits of what's to be tested
Who	key resources
When	scheduling with consideration to resource availability and project delivery plan
Why	rationale for strategy
How	what types of test are appropriate and necessary for this project and product

Test strategies are influenced by many factors, all ultimately representing risk in one form or another. As you define your strategy consider how you will account for:

- client/user expectations,
- experience and skill of the project team,
- project visibility and risk,
- project size, complexity, scope,
- quality of up front requirements definition,
- terms of the SOW (Statement of Work) or contract related to the test approach and the number of test cycles,
- project governance requirements, and
- project timelines and budget.

Just as every project is unique, so every test strategy will be uniquely tailored to the needs, priorities and other characteristics of the project in hand.

Test Plans

These are detailed plans for each type of test identified in the test strategy. Good test plans, properly executed, can make the difference between a product that runs almost perfectly for years, and one that is a constant source of missed schedules, faulty output and repeated maintenance.

Critical to the test plan will be the identification of specific known or suspected risk areas. Subject Matter Experts can provide clear guidance into these items and should be involved in both the definition of tests and the analysis of the results before a project is declared fit.

Test plans should include:

- Purpose(s)
- Approach
  - tools/medium
  - specific environment in which testing will occur
- Requirements documents or other criteria that will be used
- What is to be tested (e.g. the product itself, adjacent up/downstream products)
- Roles and responsibilities
- Testing protocol
  - testing types and phases
  - tools and methods
- Test case definition
  - purpose(s) - could be multiple in one test case
  - traceability to requirements and vice versa – every requirement should be tested, and every test should speak to one or more requirements
  - expected outcome(s)
- Test criteria (what's success?) and measurement (capturing and assessing the results)
- Data required: data acquisition and preparation can be complex and difficult, as there are often privacy and security concerns, and the manpower required to fabricate large volumes can be costly.
  - type, specific cases (e.g. empty files, invalid/missing fields/records, min/max values, fonts and languages)
  - volume
  - source/transmission requirements

Despite its length, this list is not comprehensive. Every environment has unique touch points which should be taken into consideration.

The Team

The team make up will be different for every project and will change as testing progresses through its various stages. Core team members should be involved in all steps. They include:

- project manager,
- business and systems analysts, and developers/engineers, and
- dedicated test team members if the organisation is structured that way.

The broader team will include one or more of the following:

- Subject Matter Experts (SMEs)
- Legal, Compliance, Internal Audit
- QA team
- Production/Operations (data centre, print room, call centre, distribution, other)
- Suppliers: design firm, professional services provider, hardware or software vendors, offshore development and testing resources
- Client or user

Test Execution

Test execution consists of a number of sequential activities, iterated for each test:

- execute a test,
- document results,
- confirm expected outcome,
- document any variances,
- identify defects and assign them to the product development team,
- identify changes in requirements and refer them to the project management team,
- retest, and
- report overall results.

Success Factors

Above all else, be methodical! The following guidelines are considered best practices in executing a testing protocol.

1. Define roles clearly:
    - Who is testing what?

- Who is collating results?
- Who is verifying the results?
- Who holds accountability?

2. Define the plan:
   - Don't leave it until the end of build
   - Pay attention to data
   - Get all stakeholders engaged early.
3. Set up a defect management process that includes:
   - An effective tool (and owner) for recording all tests planned and executed, and for capturing defects
   - Sample data for developers to recreate and re-test a defect
   - Traceability to requirements
   - Effective triage of defects so that major problems, and those early on in the workflow, are tackled first.
4. Understand the unique IT/print combination.
   - This is not typical application or software development
   - People who understand and are well trained in transaction driven projects are essential to the test team
   - An understanding of the architecture and the overall process is key.
5. Clear test strategies:
   - Be very clear what you are testing and how, at each phase – and what method/ medium you are using
   - Be equally clear about what you are NOT testing at each phase
   - Manage time and cost components
   - Decide how much you test and manage to that plan.

6. Use the right medium/tool for each output test. The choice of the correct medium for viewing your test output can shorten test cycle times, facilitate effective evaluation, and help avoid embarrassing errors.

- Use the target production printer or its backup
  - For accurate color, shading, image, resolution, anti-aliasing and other look and feel tests
  - Where absolute positioning accuracy is key to the test: production printers do not shrink the page
    - address block
    - barcode
  - For MICR signal testing

  - For print workflow testing including the RIP
  - For production testing
- Use a desktop printer
  - When production print capabilities aren't being tested, but content is
  - When the final output may be printed on a desktop printer, such as call center reprints from e-presentation or archive
- Use the screen
  - When print is not being tested, but content/business rules is
  - For testing epresentation or archive view.

With regard to output testing for third-party software development, where a wide range of print–room hardware may be used, data should be printed and inserted on the target hardware that the software is intended to support. This can be a challenge given the variety of output devices available so it helps to select a baseline device that you will support and then include as many other devices as is feasible. For example, if the output type is AFP print data, the target may be an Infoprint 4100 for monochrome output and an Infoprint 5000 for colour output. This is important since it may not be possible to support all devices due to both varying implementations of a standard or architecture and proprietary extensions to a given device or architecture.

Challenges

All best efforts move the plan forward, but there will be challenges. Planning for challenges is the most efficient way to ensure that they do not stop a test plan from execution or unnecessarily delay essential testing. Some common challenges include:

1. Getting data:
   - Getting data can be challenging. There are privacy and security issues with using production data; when new data feeds are involved creating them can be time-consuming and difficult.
2. Poor requirements management:
   - Time spent gathering clear and complete requirements up front pays off at this stage. If this was not the case, plan for the impact on testing and on the project cost and timeline.
   - Often a result of inadequate client engagement, expectation management or resourcing.
   - The complexity of the testing effort is often underestimated in sizing the project.
   - Testing comes late in the project, and if there have been delays, may be rushed if the project plan does not have sufficient contingency built in.
3. Regression testing:

Regression testing relies on previously created test cases and environments, with associated good output data. These are used to validate not only the new product directly, but to determine whether previous functionality has been compromised. Challenges include:

- Lack of reliable documentation to define current functionality.
- Insufficient time allowed.

4. Test case development:
   - Test cases and conditions for business rules are often defined by the business systems analysts, as they are involved in the capture of detailed requirements and understand how the system should behave.
   - Developers will create test cases for their units of code during the program build.
   - The user acceptance test team should create (or at least define) their own test conditions independently of the development team.
   - The print and mail room should define their additional test conditions to support their validation.

Types of Testing

There are many types of testing, each with its own resource requirements and goals.

Unit Testing

- Performed by:
  - Developers
  - People configuring software
  - Engineers developing hardware
- Ensures that their code/configuration/hardware:
  - Meets design criteria
  - Behaves as intended
- Should occur in each discreet new or changed module of an overall solution to makes changes easier to regression test.
- Improves quality early ("gets the lumps out"), before more complex and more expensive later testing occurs.
- Tests business rules required for properly formatted test data. If it is not available, the developer may need to build their own data, if data feed is being changed or reformatted for example.
- Does not replace need for other phases of testing.

Integration, System, and End-to-End Testing

These tests are designed to progressively test more of the project's product and its interactions with upstream and downstream systems. These tests:

- Are performed by technical resources in the testing team.
- Install software/hardware/developed code in a test environment.

- Test that the various components of the solution such as ETL, composition, post-composition manipulation, messaging, e-presentation and archive work in an integrated environment.
- Test the scripts that make the components run and work together.
- Test reconciliation: that "data in" equals "data out" from each step.

Performance and Volume Testing

These test phases ensure that the product can execute its work in the necessary business cycle window, has sufficient resources such as disk space, and provide the opportunity for identifying hotspots in code that can be reworked for improvement. They can also extend to include:

- paper and environmental testing,
- data to post office testing, and
- parallel processing.

Hardware Testing

One of the challenges in software development is determining the minimum hardware requirements for a given software product. This can be platform dependent and is an ongoing effort as operating systems, software, and hardware continue to evolve at a rapid pace. When defining the requirements, consideration must be given to whether the host system will be dedicated or not, and if not, what other components are running there and what impact they have on system resources. Test systems that meet these requirements should be maintained for all of the forms of testing mentioned.

Regression Testing

Regression testing is to ensure that existing functionality is not lost or unintentionally changed in the course of a project that modifies an existing system, or introduces a new system into an existing environment.

In the third party software world, where there may be hundreds of implementations and uses of the product across the customer base, regression testing is particularly important. It is central to any software development lifecycle work such as defect or bug fixing, as well as to feature and functionality enhancements. Once the initial test set-ups and strategies are defined and executed, samples of "known good" output should be collected and the purpose of each test case documented. These can then be used as a benchmark for regression testing any changes to the product, be they enhancements or fixes. This, coupled with an automation strategy and suitable comparison tools, allows the development team to quickly gain feedback on their changes and any potential impact it may have on the existing customer base.

QA and/or User Acceptance Testing

User Acceptance Testing both ensures that all requirements are correctly addressed in the product and focuses on features such as document objects, SLA's (Service Level Agreements) and the T&C's (Terms and Conditions) that are included in the output for each business case.

There are different names for this phase (and more than one type of testing performed) depending on:

- organisations involved – internal and external,
- project methodology,
- who the "user" is,
- nature and complexity of the project or application, and
- if there is a QA department.

A QA team tests to the business needs by using the requirements documents. Their process should be documented, structured and traceable. Business analysts and/or the client may participate in resolving questions and analyzing potential defects.

- Translated text is often tested by the translator during business testing.
- User Acceptance Testing may include functional testing of items such as usability of interfaces, searching archives, message management and local reprint. The "users" doing testing could include:
    - call centres, branches, business or technical operations, technical support,
    - marketing or product,
    - translation,
    - legal and compliance, and
    - interested third parties – financial advisors, brokers.

The ultimate user of a statement may be the account holder, but they typically only participate, as a group, in consumer focus groups and of course as individuals through industry surveys like JD Power and by way of complaints or accolades.

Parallel Testing

Not a routine test, but used when the product replaces all or part of an existing product and comparing the outputs is considered vital. It's often used for a high visibility/high risk product. It's generally run as part of QA/UAT, and compares current production output to future test output and analyzes/ resolves differences.

It may also be done as part of launch, comparing old and new output running in the production environment before the new output is delivered (and the "old" destroyed!). Typical issues:

- The request for parallel testing may be indicative of a lack of confidence in the up-front business requirements – plan accordingly.
- Parallel testing is often a manual process, takes up time and costly resources, and is prone to error; manage the volumes involved to minimize the impact.
- Be alert to privacy and confidentiality issues if live data or production output is used.

- True comparison of output requires the same input data: source and content may differ from old and new, and the data may not be available in both environments.
- The purpose of parallel testing must be clear; it cannot be used to compare look and feel if the printer is different. Understanding of the data, technology and environmental differences is key.

Production Testing

Production may be internal (in-house, or in-plant, facility) or outsourced to a service provider, who may be providing anything from print-ready to full development (raw data) services, as well as from print-and-mail only to multi-channel delivery. The testing process will differ, and it's important to communicate well, plan early and involve the right people. Production testing can include:

- color selection and reproduction,
- shading, formatting, fonts,
- graphics/pictures,
- pie charts,
- opacity of the paper,
- look and feel,
- process control information flow,
- correct process control and integrity marking,
- ADF workflow integration of the new or changed product, and
- printing, insertion and other handling or finishing.

Work Objects

Typical work objects (components) for the Test phase include:

- test strategy,
- test plan,
- test case plan,
- test reports,
- System Integration Test (SIT) plan,
- User Acceptance Test (UAT) plan,
- Client Acceptance Test (CAT) plan, and
- Production Acceptance Test (PAT) plan.

What is the End Product of Testing?

The end product of a Test is typically a test case document or library where test case documents provide proof that all tests have been met successfully, or that all major criteria have been met, and:

- there is an agreed plan to resolve minor problems,
- there are acceptable temporary workarounds in place, and
- stakeholders agree that fixes for the minor problems are not required for implementation.

Approval is required by all identified project stakeholders. It is not just management, but also the teams that have operational responsibility that must approve.

Implementation Planning

The point at which implementation planning should start will vary with the type of project, but it generally needs to be complete by the end of the Test phase so that implementation can proceed in a timely fashion. By the end of Build & Unit Test the overall shape, schedules and dependencies of the new/changed computer jobs, and/or print-room workloads, workflows and equipment layouts should be clear.

The implementation plan should include a checklist with:

- required production job scheduler changes,
- JCL/scripts for executing the new/changed jobs,
- storage (disk, tape, etc.) and network bandwidth requirements,
- design for the change implementation package and process, including backout, and
- required operational documentation is required by operations.

Implementation planning should include any scheduled outages, maintenance downtime, and other known potential delays necessary to schedule and execute the changes. It is critical to take account of other impending changes, business volume peaks and troughs, and other factors so as to minimize the business impact of implementation.

Launch (Production)

Contributed by Rebecca Rodgers m-edp

What is Launch?

Launching is not a one-day event, but a culmination of successful requirements gathering, analysis, development, testing and preparation for transitioning into production.

The areas discussed in this section include the basics of what is needed to be considered early for successful production / launch in the transaction document space. The inputs and sub-components of launching a change to a new or existing project or production jobs, tracking progress, establishing procedures for back-out, and managing the Launch stage rely on the successful completion of the previous stages. Launch steps and considerations apply to all types of projects including new applications or changes, adding new or upgrading hardware, adding new or upgrading software. It takes into account all stages of preparation and production for each channel from data receipt and composition through production to delivery (print or e-presentment) and archive, and even to return functions when applicable.

Launch defines the activities to run the product / application live for the first time in production with the intent to enable ongoing production.

The following terms are commonly used as synonyms to discuss the process of launching for production:

- Going Live
- Production Step
- Promotion to production
- Hand off to Operations
- Project Close Phase
- Transition Phase

Steps in the Launch Sequence

- **Requirements Gathering:** Identify target launch date and alternative, if available.
- **Business Analysis:** Identify and detail what the product components will be in production (physical and electronic). Identify changes to current product / environment; everything from training through to materials and process. Adequate times are allocated for adding or upgrading hardware or software if the project requires it.
- **Technical Analysis:** Identify how the product will be manufactured in development, test and production (physical and electronic). Impacts to archiving and e-presentment process and systems.
- **Stakeholder Adoption:** Identify who cares and why.

- **Design:** How will the design influence manufacturing materials and environment? Will the entire design be archived or e-presented.
- **Architecture:** How will the manufacturing components work together (physical and electronic? Do the product "cohabits" satisfactorily with the existing architecture portfolio?
- **Development:** Skills, components and timeline.
- **Testing and Quality::** Skills, components and timeline including all channels (print, e-presentment, SMS, voice, social media).
- **Production (aka Launch):** All previous steps are complete, final readiness and go-live is on track.
- **Maintenance:** Maintain what went into production (accuracy, quality and on-time) for all channels.
- **Decommission:** It must be possible to undo the entire production without negatively impacting anything else.

Preparing for Launch

The key activities surrounding the Launch include:

- communication,
- staff training,
- management of process,
- staffing and personnel management,
- logistics planning,
- paper and environmental monitoring, and
- Service Level management.

Communication

There are many participants, constituents, and stakeholders in the communication workflow which require a wide range of communications and coordination requirements. The " who, how, when, what, and why" all need to be planned for ahead of Launch. Communication goals must include that every department is fully informed of Launch. This includes ensuring that they are fully informed of the intentions and consequences of the launch as well as their role and responsibility during pre-launch, launch, post launch and production.

Speaking the same language from project inception is key to the success of the Launch. Confirming that you will be gathering data and information using the same metrics like impressions (versus sheets or pages) from start to finish will ensure accurate tracking and billing. For example if you bill by impression then you must start tracking impressions as early as file receipt (print ready files) or composition and keep track of them through all processes. Once the impressions have been assembled if you make changes to the impression count in a post-composition process it must be recalculated and tracked as changed.

Some applications require special tracking for billing. Manual jobs or jobs with unique client characteristics can cause major delays if they are not identified and tested during Launch. The testing has to include all facets of the jobs, all of the way through to the billing. Successful Launch assumes you have done everything thoroughly for all variations and communicated well with every participant and stakeholder before and during the launch stage and also immediately following launching into production to ensure a clean hand-off.

Failure to Launch happens for many reasons. It often means someone didn't know they were supposed to do something or the timing wasn't clear. Sometimes key items are missed because they were not written down or scheduled. Tasks like using a promote code, executing an alternative script, processes for ordering /receiving / processing stock can cause delays if not including in the Launch plan. Similar problems occur when deadlines are overlooked, data is not received on time and there is no process to manage the delay, when output is not held (a parallel run was not accommodated properly and the wrong output is mailed), wrong stock is used (no plan for business continuity or back out). However, with a formal Launch process it is easier to determine the issue and fix it by examining the audit logs or Launch documentation

Training

Launch requires that everyone involved in handling the product in production receives appropriate training. This includes operators (all shifts), production coordinators, data center teams, inventory management teams, web team/s (internal and external portals and archives) and support groups like production support , call centre, billing groups, escalation groups, and business continuity sites. There can also be special training requirements when something outside of the automated / normal process requires special handling. This generally entails manual handling and tracking for integrity and billing purposes.

Management of Process

Launch requires knowing how to manage the project lifecycle going into the launch phase and subsequently managing the product in production. Pre-launch and post-launch management may incorporate different tasks depending on the project plan and organizations launch process, but Management of the product and process does not stop once the project has launched.

Moving into production should engage rigorous policies and procedures to ensure that:

- the job scheduler is updated to run automatically or as needed,
- new or upgraded hardware or software is completed,
- back-out procedures have been documented and communicated,
- the dataset space is properly allocated and accounted for,
- ABEND (abnormal end) instructions are available,
- code freezes have been addressed,
- 3rd shift production teams know who to call with problem, and
- any unique processes are documented and well understood.

For example, when a newly composed statement package is moved into production via a formal process:

- object code itself is promoted,
- resources used are promoted,
- any new emitters are promoted,
- scripts (or JCL) used to run job are promoted,
- tracking process, data/codes etc. are running to capture billing and integrity reporting information, and
- entries in a scheduler to kick off the jobs have been promoted (script or JCL):
  - batch process,
  - manual.

Using a formal process provides you with the advantage of knowing if anything has gone wrong and with the ability to 'back-out' if needed.

Launch objectives are typically two-fold:

1. Document the software and/or hardware (and results) as it goes into regular production.
2. Every department is fully informed of the launch and what they need to know or do.

Post-launch management of the product relies on proven and consistent processes on and off the production floor including but not limited to:

- inventory management,
- workflow management,
- tracking and reporting,
- environment management,
- supplier support, and
- staff management.

Staffing and Personnel Management

Moving a product from a project lifecycle into a production lifecycle not only requires training and communication, but often disrupts normal work cycles. There is often a requirement for a subject matter expert to be available in their off hours for extended periods, therefore staff management has to take into account how Launch and immediate post-Launch will be manned.

Staffing for Launch can incorporate weeks of preparation, including training, testing, equipment moves and general preparation. Staffing appropriately for the preparation must take into account that the actual launch event is also staffed appropriately.

Personnel management includes, but is not limited to:

- managing inaccurate / incomplete tasks or assignments which can result in service level penalties,
- rewards and recognition, and
- career advancement and succession planning.

Logistics Planning

There are many tasks in launch, most of which must be identified and planned for in advance. Identifying the tasks, planning for the complexities of production and implementation, and controlling the flow of production (the manufacturing process) across all delivery channels requires planning and testing. Considerations include, but are not limited to:

- adequate lead and execution time allowed for adding or upgrading hardware or software,
- adequate lead and execution time allowed for system / network environment changes,
- the support groups can manage product nuances / changes reliably,
- ensuring that the files can be received and run at the expected time,
- ensuring enough IT capacity (processor, memory, disk, bandwidth etc.) to handle the workload at the expected time,
- the print and finishing floor can manage and process the output,
- production floor has required room to receive and process product,
- environment is managed to support paper types and forms working reliably,
- integrity checks work at each manufacturing step,
- rerun/reprint facilities are in place and tested,
- suppliers can deliver as required,
- warehouse can handle additional inventory volumes as required, and
- archive and e-presentment environments can process and present at the expected times.

Paper and the Manufacturing Environment

It is important to plan for the handling of the paper products for implementation and ongoing production. These products include all the substrates for the document set created on the manufacturing floor, including paper (roll, cut sheet, or fanfold), inserts, tip-on material, outer envelopes (OE) and business reply envelopes (BRE).

For Launch, the previously approved products be received in advance, unpacked, checked and validated as correct and in good order. If they are accepted they must be entered into inventory, acclimatized and properly prepared for production. The final step will be to move them to the production floor ready for the manufacturing run.

Monitoring the manufacturing environment for production is an important part of everyday planning. The environment should be adequately vented, heated, and humidity-controlled to accommodate both the substrates and the equipment according to both the vendor/supplier specifications and to the fluctuating conditions of the location. Open delivery doors, improper humidity levels, heat and cold all impact how well the paper behaves during printing and finishing.

Electronic document environments and paths must also be considered. Ensuring that all files are received with ample time and space to prepare and consolidate data/files into the finished product. Check temporary storage levels and file security settings as part of the checklist.

Service Level Management

There are many elements to managing the manufacturing processes to maintain contracted / expected service level agreements (SLAs). Key to the process is the management of the metrics to ensure that SLAs are met are and the metrics required for management and client reporting are gathered and forwarded.

Metrics for tracking on-time delivery and quality are two of the most common requirements. If the manufacturing environment is not set up at an Automated Document Factory then there are many pieces of paperwork and many sign offs that follow the product as it progresses along the manufacturing workflow, usually in the form of a job docket/ticket.

Capturing these same metrics is also required for electronic delivery ensuring the full (expected) electronic lifecycle is achieved. It begins with the successful manufacturing of the files, initial delivery to consumer, delivery to call center/agent, delivery to archive, and the copy to the backup location at a minimum.

Launch Infrastructure

Some of the tools for Launch consist of:

Strong project team management, for example:

- Good communication, appropriate approvals, failover, documented back-out, appropriate tracking, status reporting.

Good documentation:

- Pre-launch documents will consist of a variety of sign off/ acceptance documents (test, materials, training, operations, executive, client to name a few), new / updated process documents, vendor/supplier approvals.
- Post-launch documents will consist of operations sign off, client sign off, post implementation review brief, executive sign off to name a few.

Access to support:

- Knowledgeable skilled support staff, appropriate hours of availability, adequate authority to make decisions.

Product Integrity and Project Integrity

Product integrity refers to the quality of the output (paper or electronic). The documents intended for the recipients (and only those documents) are present either in the envelope or electronically available, the document set is complete, accurate and of good quality and the document package is delivered / presented on time.

Project integrity is not only specific to the Launch, but covers all steps leading up to Launch, and the post implementation steps. It also refers to how well the project is managed, communicated, documented, and implemented.

Key Stakeholders

Launch requires coordination with more departments than any other phase; other departments may be directly / indirectly affected by launch or during launch phase and consideration of the impact to each of them is crucial.

Notification and approval for Launch should be accomplished with formal meetings, not just by email.

Nearly all stakeholders have a requirement (or interest) to sign off. The Launch can be killed by any key stakeholder if something goes wrong or is perceived to be wrong to the point that it will affect the outcome (e.g. operations management, compliance / risk, document owner/s, technology, business sponsor).

Vendors do not usually promote or Launch, however they can play a part if they have been contracted to have a role. Even facilities with long term contractors usually use only employees for formal Launch processes.

Key staff should understand the new or changed product or service in order to train:

Ongoing trainers (to cover for absentees, career changes etc.)

- sales people,
- call centers, and
- operations.

Stakeholders will include, but are not limited to:

- **Quality Assurance (QA):** Generally defines the activities conducted throughput the project (especially the testing phases which are usually a subset of QA) to ensure accuracy and completeness of project execution. QA sign off is a precursor to Launch.
- **Sales** may have been involved throughout the project or may need to be responsible for client care during Launch.
- **Call center / customer service:** Must have received notification / training on the changes and may even have signed off on some business requirements or goals.
- **Call center / order entry:** Must have received notification / training on the changes.
- **Web services:** Electronic launch needs to be appropriately coordinated with physical Launch.
- **Upstream departments** (e.g. DBAs, support, Legal/Compliance, etc.) must receive notification / training on the changes.

- **Downstream departments** (mail room, billing, reporting etc.) must receive notification / training on the changes.
- **Operations** (e.g. print room, finishing, distribution, support) must receive notification / training on the changes and likely have signed off on some business requirements or goals.
- **Disaster recovery/Business continuity** must receive notification / training on the changes.
- **Training:** On standby for support.
- **PR department:** Announcements need to be appropriately coordinated.

Launch is not an end step in the document lifecycle, but it is usually a well-defined one indicated by successful output, client and operations sign off. Successful Launch is worth celebrating.

References

Some materials are courtesy of acadami™ course materials

Maintenance

Contributed by Franklin Friedmann edp

What is Maintenance?

Maintenance as a process represents a continuing activity to sustain the production system throughout its operational life, covering update through decommissioning. Maintenance activities include those necessary to operate, correct faults, and address enhancements in a production system that includes hardware, application software, proprietary software or a combination. It also includes the documentation of these processes.

Processes

An operational environment consists of all elements that ensure the smooth running of all business processes. These processes and their maintenance involve all software application and their supporting structure, which may be computerized or human. Included in maintenance elements are:

- The application, including operational code, documentation of processes, instructions to operators with error conditions.
- Operational staff that may need to take an active role in starting applications, as opposed to automated start, or provide responses that cannot be automated.
- Supporting staff who responds to questions and problems. This is likely to be a layered approach, with a Help Desk in their predefined role as the first line of support performing triage. Secondary support may be the staff with access to documentation to explain the problem. If they cannot respond, a third level includes application specialists and developers.
- Hardware, including printers, inserters, finishing and mailing equipment present a more nuanced problem. They represent mechanical devices requiring regular maintenance beyond the remainder of the Information Technology universe. Where some devices have an MTBF, Mean Time Between Failures, that is measured in years and may last longer than the expected utilization of that device, mechanical devices usually require attention based on usage and time.

The Lifecycle Effect

Maintenance requirements, while always present, are not static. Applications, software, and the business processes that support them move in lifecycles, some longer than others. To ensure that all maintenance requirements are met the business objectives for maintenance operations must be clear. Attaining the business objectives is more important than delivering every technical function, but it is necessary to include the following items in maintenance objectives:

- The project parameters – longevity, start and end dates.
- The application and archiving strategy, which must include what data retrieval is required (not necessarily caused by a disaster).
- Methods for continuing upgrade and correction.

- The standards that would set a future disposition date.

The maintenance process must be structured to ensure that every facet is documented and managed. This includes:

- Change requests and controls, covering:
    - the system (with underlying software and hardware),
    - the applications, and
    - the documentation that will be updated.
- Change testing with incorporation into the operation.

While in theory every process should work perfectly, in reality controls are required to ensure that maintenance is viable. Common controls include:

- Version controls – particularly where multiple versions are in use simultaneously, either on the same or separate platforms, with their differences identified and maintained.
- Usage transition controls between versions, so that older versions can be decommissioned.
- Compatibility – forward and backward. While newer versions may have additional capabilities. Though upward compatibility would seem reasonable, there should be no assumption that all the capabilities in an older version will be ported to the new one; upward and backward compatibility must be verified. In addition, no assumption can be made that new functionality will be retrofitted to earlier versions.
- Compatibility of all facilities carried upward with the older version.
- Help support to cover all versions in operation.

When Does Maintenance Happen?

Maintenance may be required for many reasons, from a failure in a specific point in the process to undefined failures. To ensure acceptable performance, operations may require maintenance on any single process or group of linked processes on a regular basis. Separately, performance systems may have maintenance points designed into the Lifecycle. Those maintenance points may be called out in a response or process delivery agreement such as a Service Level Agreement (SLA).

To perform an analysis of identified problems that trigger maintenance operations, especially when a problem is not classified as a true failure, requires some additional tools. They include:

- System Metrics and Key Performance Indicators selected by the installation to represent the performance characteristics most important to management. These may be generated automatically based on the more detailed measurements below or prepared as part of a summarized report.
- Support monitoring utilities to measure operational performance. Additional equipment based on the results will result in a new set of measurements. With measurement, load balancing is possible.
- For processors this usually represents utilization and response time.

- For printers this represents up time, utilization and throughput,
- For ancillary devices such as storage mechanisms, traffic (quantity and size) and response time are measured.
- System Availability reports provide the usable hours, identifying failures and repair activity. They may also provide business metrics such as time from job arrival to delivery to its final process.
- Job tracking, audit and reconciliation should be addressable from the statistics.
- Process defects should be identified and followed, usually in a report that lists these problems with a target resolution date.
- Review meetings with suppliers' service organizations to address any problem and performance issues.
- A report similar to the defect entries to inform follow-up meetings.
- Recommendations.

Documentation Support

Documentation Maintenance must be part of any operational plan.

- Developing and updating documentation in a reliable manner, ensuring that changes to the system or application are entirely mirrored by the documentation at time of implementation is essential for proper operation.
- Ensuring that the documentation accessed matches the version in operation for each environment. Where multiple versions are active, identification is essential.
- Ensure this documentation is available to the Help Desk and searchable.

Problem Resolution Support

Support must be part of any operational plan. Designated support normally takes the form of a Help Desk operation with access to all required environmental information and documentation, plus a track of past problems. Once resolved, each problem and resolution must be documented, available for review by the Help Desk staff, and usually review by the users. Expert advice should be available on call.

Training Support

Training and scheduling to support the changes must be put in place.

- A prerequisite: maintenance must assume that users attain self-sufficiency; therefore a training plan must be implemented before going live.
- Quality documentation allowing trainers to be self-sufficient to perform knowledge transfer improves the prospect of success.
- New business opportunities and/or issues must be documented as applications go into production.
- The operational environment must be prepared for new applications.
- The Help Desk must have a process for resolution.

Equipment Maintenance

Equipment Maintenance is distinct category. It is the glue to the entire operation; it is also the one area that management will immediately trace back to when a failure occurs. It is often contracted to the manufacturer, supplier, or certified 3rd party.

The class of maintenance depends on the device:

- **On failure:** Relates to devices with a Mean Time to Failure (MTTF) that exceeds the useful life of the device.
- **Preventive Maintenance:** Scheduled actions based on usage and/or elapsed time. This equates to the actions that car manufacturers recommend for your automobile. Today, production printers must be in this category.
- **Predictive Maintenance:** The most recent approach, it requires evidence gathering and sophisticated assessment to determine when to take action. This is more likely to be used in factory maintenance of critical machinery.

There are known strategies to avoid disruption:

- Assume there is sufficient equipment during planned maintenance, the strategies reflect unexpected outages.
- Failover or fallback – a planned procedure that supports the use of secondary equipment. An essential component is continuing access to the data to ensure continued operation.
- Secondary equipment may work at a degraded level. The decision to use this path is based on the economics of a possible outage.
- How much will it cost in lost revenue, damage to the image of the corporation or government and the cost to recover the operation?

Cost management and budgeting is an integral part of the tactical and strategic activity to ensure the operation meets the goals set out by management.

- To bill users you need the ability to track usage, agreed upon standards and where possible, incentives to drive workloads to light periods. This approach should address Service Level Agreements (SLA), which should pay a premium for guaranteed service.
- In print, supplies create an additional component. Whether it is the printer consumables such as inks, toners or ribbons, or the distribution elements such as paper, envelopes, inserts and the postage process, these costs represent a major expenditure that require detailed accounting and attention.
- Warehousing the supplies and distribution elements is a major activity, often taking space that is multiple times the operational printer area.

Software and Application Maintenance

Upgrade of the processes is a recurrent activity. Whether for the underlying infrastructure or the business application, the ability to make changes without impacting production is a must. To ensure this is possible there should be a test environment mirrored on the production that can simulate the operations.

- Changes range from application updates to regulatory changes that have mandated “go live” dates.
- In print and on the web, updates may address marketing tailored to customers. These include adding color, messaging, forms replacement and rebranding.
- Tests range from unit/component modification, which will compare the current activity against changes through a full test. A full test must cover anticipated volumes.
- Stress tests are essential where changes could generate loads for which the environment was not designed.

Improvements in production processes should be part of maintenance:

- automation of various steps currently performed manually,
- *continuous improvement*, *lean* or Six Sigma goals to eliminate waste and improve efficiency respectively, and
- adding controls to print output such as bar codes for integrity improvement to QR codes to encourage customer action.

Decommissioning

Decommissioning an application or system should invoke a formal process that must cover:

- the organization’s current needs, usually set by Executives,
- planning for the possible disruption to current users on the system being decommissioned,
- where existing capabilities will be handled; some users may lose flexibility such that they demand the continuation of the older system/version,
- continuation - a costly process with duplication not only in staffing but possibly in data, which is generally unacceptable,
- recognition and response to relevant legal statutes to ensure that the data and access is maintained through the normal end of life should there be active litigation,
- housekeeping for all administrative activities included in the maintenance of the system being decommissioned,
- necessary material removal,
- cancellation of access,
- unique items, particularly in legacy applications that were not subject to current standards during development; a particular consideration that arises is the staff that built the system without full documentation standards in place may have left the operation,
- transition to the new environment / system; this new system may be an existing application that was run in parallel,
- data management and transfer,

- assurance that users can transfer to the new environment with procedures to assist the transfer, and
- documentation relating to the transfer provided.

Document Production Workflow

Data

Contributed by Kevin Tondreau edp

What is Data?

The term data means different things to different people. What does it mean in the document creation world? What types of data can be presented to document composition tools or to programmers to code and design a document? What type of data is required for printers to perform their magic to generate a printed page, or for an online archive solution to provide internal or external document access? What are data types and how are they generated, stored and used?

Raw data (a transaction document systems phrase, meaning data in its native format in customer databases, accounting systems, and so on) is composed of records, fields within the records, and characters and numeric values within the fields. How the data is analyzed and formatted depends on its role. For it to be parsed and presented correctly in its final format requires a basic knowledge of data structures and correct data storage protocols.

Languages of Data

There were originally two basic language types used to store data: EBCDIC (Extended Binary Coded Decimal Interchange Code) and ASCII (American Standard Code for Information Interchange). To support extended character sets, Unicode has emerged as the standard encoding system for a majority of computer systems. Unicode defines a code space of 1,114,112 code points, each available to map to a character.

Each language uses a different approach to describing data. Basic character mapping is accomplished through the use of code pages that map code points to hex values that are interpreted based on the language of the encoding.

EBCDIC is the common language for IBM Mainframe Operating Systems. It describes letters, numbers and standard punctuation marks. EBCDIC evolved to an expanded character set up to 256 characters using an eight-bit encoding schema.

ASCII is used in Windows, UNIX and most non-mainframe systems. ASCII contains letters, numbers and standard punctuation marks. Depending on the system storing the data, the standard seven-bit string allows 128 characters; some systems will expand to eight bits of data, which allows for an expanded character set of 256 characters.

Systems that permit 32 bits of data allow for characters in many of the world's languages that require more than 256 characters to be encoded into a single code page.

Unicode is emerging as the preferred encoding system to allow multi languages to co-exist in a single document regardless of the presentation platform (print or view). This has greatly reduced the errors and incompatibility in displaying and transforming information.

There can be some issues in translating between or among the three schemes since there may be mismatches in the number of characters supported. For a standard translation of the numbers 1, 2 and 3 the Hex values are:

```
EBCDIC -    X'31', X'32', X'33'
ASCII  -    X'F1', X'F2', X'F3'
Unicode -   X'0031', X'0032', X'0033'
```

Packed Decimal Format

Packed Decimal Data format was developed by IBM in the 1960s to increase transmission times and reduce overall storage space. Packed-decimal format means that each byte of storage (except for the low order byte) can contain two decimal numbers. The low-order byte contains one digit in the leftmost portion and the sign (positive or negative) in the rightmost portion. The standard signs are used: hexadecimal F or C for positive numbers and hexadecimal D for negative numbers. The packed-decimal format looks like this:

Decimal	BCD
Digit	8 4 2 1
0	0 0 0 0
1	0 0 0 1
2	0 0 1 0
3	0 0 1 1
4	0 1 0 0
5	0 1 0 1
6	0 1 1 0
7	0 1 1 1
8	1 0 0 0
9	1 0 0 1

The packed data format reduces the storage space by half. For the digits 92 in a longer number, for example:

Standard 8 Byte Storage

```
Decimal         9         2
Binary 0000 1001 0000 0010
```

Packed Data Storage

```
Decimal         9    2
Binary       1001 0010
```

Packed data format also needs to assign a positive or negative value.

Sign	BCD	Sign
Digit	8 4 2 1	
A	1 0 1 0	+
B	1 0 1 1	-
C	1 1 0 0	+
D	1 1 0 1	-
E	1 1 1 0	+
F	1 1 1 1	+

The value of -1,234,567 is encoded as:

```
0001 0010 0011 0100 0101 0110 0111 1101
   1    2    3    4    5    6    7    -
```

Data File Structure

Data file structure can take on many different variations. It defines methods to store information and use the information to generate a readable document. Key to this is the use of metadata to give each field meaning. In the Fixed Length Records and Fields example below the column headings (First Name, Last Name etc.) are metadata that give meaning to the data values. The different common file structures described below each have the metadata conveyed in their own way.

Fixed Length Records and Fields are common. They have a defined length for each field and a maximum length of all fields combined. In the example below the overall record length is 67. Each field can only hold the maximum number of characters defined in the record layout.

	First Name	Last Name	Address	City	State
Length	10	15	20	20	2
Position	1-10	11-25	26-45	46-65	66-67

Variable Length Record and Fields are also common in the print and processing area. Using variable length records saves disk space. There are record descriptors in the front of every record to allow the program or user to know the correct length of the record. The most common record descriptor is a 4-Byte RDW (Record Descriptor Word). Interpretation of RDWs is very basic. The first two bytes define the overall length of every record.

A simple example, where the value in the file is a HEX X'2600', this value in decimal for the first two bytes and record length would be 38 positions long. The last two positions are reserved and not used to determine the overall record length.

Delimited Fields are common for storing and reading data. A delimiter is used to separate fields within the same record.

An example of a layout for name and address using a basic delimited file would be (Sam,Smith,15 Main Street,Anytown,MA). Here a comma is used to separate the fields. There can be a risk in using a comma to delimit fields. Some fields can contain a comma as part of the field. A person's surname can contain a title (Smith, MD). Here the value MD would be separated into its own field. A common approach is to place each field within or choose a delimiter that will never be part of the actual fields. All are valid formats and can be used to parse and use the data.

```
(Sam,Smith,15 Main Street,Anytown,MA)
('Sam','Smith, MD','15 Main Street','Anytown','MA')
(Sam|Smith, MD|15 Main Street|Anytown|MA)
('Sam'|'Smith, MD'|'15 Main Street'|'Anytown'|'MA')
```

XML Data Structure Records are a set of rules for encoding documents for data presentment. The record format is both human readable and computer readable. It is broken down into a few elements identified by tags identified by angle brackets. These tags are also the metadata. Each tag must have an open and close and within the tag there may be elements which also require and open and close tag pairs. In the example below there is a document with transaction elements and summary information.

```
<document>
<MailState>NEW YORK NY 10019 -</MailState>
<FeesCharged>$0.00</FeesCharged>
<InterestCharged>$0.00</InterestCharged>
<openclosedate>01/05/12 - 02/04/12</openclosedate>
<TRANS>
<transdate></transdate>
<transdescr>PAYMENTS AND OTHER CREDITS</transdescr>
<transamt></transamt>
</TRANS>
<TRANS>
<transdate>01/05</transdate>
<transdescr>Payment - Thank You</transdescr>
<transamt>-1,000.00</transamt>
</TRANS>
<TRANS>
<transdate>01/08</transdate>
<transdescr>Payment - Thank You</transdescr>
<transamt>-1,000.00</transamt>
</TRANS>
 <TotalFeesChargedInYYYY>$0.00</TotalFeesChargedInYYYY>
<BALGRP>
<balancetype>Purchases</balancetype>
<apr>0.00</apr>
<adb>$1,690.42</adb>
<finchrg>-0-</finchrg>
</BALGRP>
<DaysInCycle>31</DaysInCycle>
</document>
```

Record termination allows the program or operating system to know when a line feed (new line) is needed for presentment and readability. Windows primarily uses a carriage return and a line feed (Hexadecimal of X'ODOA'), whereas the UNIX environments use a single value X'OA' to terminate a line.

Multi-Record Format Data describes a file constructed of data for processing by defining the record layout within a file by the record type.

```
12345678901234567890123456789012345678901234567890123456789012
010NAME     ADDRESS1    ADDRESS2     CITY     ST     ZIP    ACCTNUM
020DATE   TOTALDUE     INTEREST    PASTDUE      DATEDUE
030DATE DETAILTRANSACTION                    AMOUNT     INTEREST
030DATE DETAILTRANSACTION                    AMOUNT     INTEREST
030DATE DETAILTRANSACTION                    AMOUNT     INTEREST
030DATE DETAILTRANSACTION                    AMOUNT     INTEREST
990ENDOFRECORD
```

.In this data type the record type in the first few bytes of the record determines how to read each record for processing. Each record type will have a different position and different fields pertaining to the records. In the example above, the 01 record is the header record and all records between 01 and 99 are tied to the same recipient. Here each type of record (01, 02, 03...etc.) will be used to generate a document for print or Web view.

Field storage and extraction for processing are different from system to system. A relational database could have many information tables. Storing dates, currency, and names with upper and lower case sensitivity can require different formats. Presenting the date and currency in the correct format starts with how the information is stored and what the final intended presentation format should be. Most data element storage eliminates special characters and punctuation. A dollar value of $1,250.55 would be presented in a data file as 125055. This allows the presentment of text based on the country and separate placement of the currency character. Dates are more complex and are stored in many variations. A common format is to use a two position month and two position day of the month, with a four position year: 12012013. This allows the end presentment to be displayed in accordance with local or user conventions. This conversion typically takes place in a process described in the following section.

EXTRACT, TRANSLATE and LOAD (ETL)

Extract, Translate and Load (ETL, also called Extract, *Transform* and Load) is the process of gathering data from various source systems and preparing it for use.

Extract: Gathers the data from its various sources

Translate: Normalizes it by converting it from its stored forms into consistent human readable character data, and formatting the character data (date, currency, phone numbers, etc.) to conform to national or other conventions.

Load: Writes it to suitable files and locations for the composition step.

Most modern composition tools contain some ETL functionality, but given the variety of database systems in use and the complexity of most business processing schedules, the most efficient and effective ETL will generally be achieved by custom programming work in-house.

The date is an example which can be stored and presented many ways depending on the country of origin. The format MM/DD/YY is common to the United States. Most of Europe uses DD/MM/YY. Japan uses YY/MM/DD. The separators may be slashes, dashes or periods. Some locales print leading zeroes, others suppress them. If a native Japanese speaker is reading a US English web page from a web site in Germany that contains the date 03/04/02 how do they interpret it?

ISO 8601 specifies YYYY-MM-DD, 2003-11-05. Looking at 11/05/03 on a document a US citizen would read as November 5, 2003 and some in Japan would see May 3, 2011.

Below is a record from the USPS that contains a scan of a mailing document within the USPS process. Highlighted in bold is the date including a time stamp for when the document was scanned at the USPS sort facility.

21280,891,**09/20/2013 23:59:59,**80919123412,00050021280300000060

30026,919,**03/08/2013 05:36:44,**30047430337,00270121280300000061

The programmer must define the record layout of each field for processing. This record contains five fields which are described below. The task of the developer is to determine the layout of the date and how to parse the information and store the data. Below is a snippet of code which outlines the format of the date field.

```
FILE NAME=Description='OneCode Confirm data file',FORMAT=CSV,
  FIELD=(uspszip,1),
  FIELD=(opcode,2),
  FIELD=(datetime,3),
  FIELD=(userzip,4),
  FIELD=(uniqueid,5);
```

The date format the developer would present on a document as:

September 20, 2013 11:59PM and March 8, 2013 5:36AM

Might be coded as:

```
DOCTRACKINGDBSESSIONINIT=alter session set NLS_DATE_FORMAT =
'yyyy-mm-dd';alter session set NLS_TIMESTAMP_FORMAT = 'yyyy-
mm-dd HH24:MI:SS'
```

The process of storing date information as numeric values allows the processing environment to sort effectively and quickly. Storing months by name does not allow correct sort routines.

Currency storage can be basic, but presentment is the challenge. Storing the value as numeric only has its advantages, but the final presentment requires a process. Storage of the value of 1,234.99 would be 123499 and -1,234.99 would be store as 123499-. Presenting the amount depends on the origin or the consumer. The developer would be required to format the data and place commas and decimals when required. The dollar value above can be presented as $1,234.99 where as the negative value can be presented as:

```
$1,234.99-
-$1,234.99
($1,234.99)
```

Different currencies have different formatting practices, for example separating thousands with commas or periods, and some have different numbers of decimal places.

Similar conventions apply to telephone numbers in different countries (e.g. France, where numbers are presented 0A.BB.BB.BB.BB), and to other data types (e.g. social security and health insurance numbers), depending on the business application area.

Document Objects

Contributed by William Broddy m-edp

What are Document Objects?

When we compose a series of documents, we use document objects within the document(s) to create:

- typography,
- barcodes,
- business graphs,
- photos and scanned images,
- text blocks and tables, and
- color information.

There are three general categories for document objects: vector graphics, pre-rasterized images, and compound objects.

Vector Graphics: A series of instructions describing the object with draw and fill commands. Typographic fonts, barcodes, and business graphs are usually described in this way.

Pre-rasterized Images: A matrix of pixel-points that describes where to place black dots, grey-shaded dots, or color dots. Photographs and scanned images are the most common types of raster objects. However, legacy applications may still use pre-rasterized images for logos or fonts.

Figure 1 - Comparison of pre-rasterized vs. vector objects. http://en.wikipedia.org/wiki/File:Bitmap_VS_SVG.svg

Compound Objects: A series of layout commands and objects (or references to objects) that define a table, mailing address block, or message. We use compound objects to define a discrete area (also known as zone or region) of a page or document that is independent of other areas. More sophisticated compound documents also manage color attributes and profiles for the content within the object.

Document objects are placed and customized based on business rules and by data associated with the recipient or with the type document.

There are two things required to compose a document: data that is unique to the specific document or series of documents and document objects, pre-defined parts of a document series.

Document objects are created as part of the document development process and stored in the application libraries associated with the document composition or modification process. Different object-types are used to maximize document fidelity (to ensure that the recipient views the document as it is intended to look. By using widely supported document object-types document portability is maximized. Document objects are also tuned to optimize the speed that the document can be rendered for printing, or for viewing on a screen.

In the early days of laser printing and screen display, circa 1980, virtually all document objects were stored as raster images. This was because the Raster Image Processor (RIP) could not render vector commands into raster images at a sufficient speed. Today most print manufacturers take advantage of video processors developed for video gaming which are tuned to render vector graphic objects into raster patterns very quickly. Today, any document object that can be easily described in vector commands should be created as a vector graphic object, e.g. logos, icons, fonts, barcodes.

Aliases and Terminology

Print resources: A name for legacy document objects that reside in libraries attached to the printer or print-server. These are fonts, logos, and text blocks (e.g. pagedefs and formdefs). These print resources are bound to the print file as it is sent to the printer RIP. This was historically done to minimize the size of print files when telecommunication bandwidth was very limited. Virtually all print resources are document objects. However, there are newer document objects (such as color management objects) that don't conveniently fit into the print resources category.

Zones / regions: The area in which a text block resides. For example, the address text block may also be referred to as the address zone or address region. However, an address region may not be a discrete text block; it may be line text positioned from the top left corner of the page and the text block it is within may also contain other information not associated with the address, e.g. the date of issue.

Hard vs. soft resources: The same document object is often used repetitively within a document or series of documents, e.g. the corporate logo. To eliminate re-rendering each time the object is used it may be defined as a hard resource so that it can be rendered once and then cached for repeated placement. A soft resource may be used only once, such as a text block describing a table of individual transactions for specific recipient.

Document Object Management

The development and use of document objects must be properly managed as part of the ongoing maintenance and upkeep of document applications. For example, many organizations use document objects to produce corporate logos and trademarks on their documents. Over time, instances of these objects will reside in many difference libraries or hard-coded into applications. If not properly managed, it may take months to introduce the corporate logo in all the locations it is found.

With the introduction of message management within transaction documents, we need to tightly manage the development, test and promotion of compound objects containing marketing messages. These marketing messages are often encapsulated with vector-graphics and/or photographic images, and usually include references to typographic fonts. Marketing departments typically have much shorter deadlines for launch into production than the classic IT change control process allows. Therefore, the tools that manage development, test and launch of these objects follow a different development path than other document objects.

General Object Categories

Document objects have evolved over the past 35 years since the first laser printers were introduced. The generations of specific object-types are listed in reverse chronological order.

Vector-based objects

The current generations of the following document objects are usually created using vector commands. However, there may be legacy formats still in use that date back to the early days of laser printing that may be in raster format.

Vector-based objects exploit the power of processing chips designed primarily for video game display. These chips can render vector-based objects at incredible speeds. Describing an object using vector commands provides a number of advantages:

1. The object size is usually much smaller than an identical looking object stored as a raster object.
2. The RIP processor can turn the object into a raster image that is the correct resolution and color profile for the printer (or video display) on which it's rendered, e.g. a vector font character will be converted to a 600 dpi image that can then be enhanced beyond 600 dpi through antialiasing on the specific printer. If the font character is to be printed in a specific RGB color, the RIP can tune the CMYK color percentages used based on the color profile of the printer.
3. The object can be oriented in any direction (0 – 360°) without affecting the fidelity of the object.
4. The object can be easily resized with affecting the fidelity of the object, i.e. the roughness of the edges and the graininess of the fill.

Examples of vector commands include:

- draw line,
- draw circle,
- draw box,
- draw ellipse, and
- fill.

Historically, vector-based objects were introduced in the desktop publishing environment by Apple and Adobe. The speed of the RIP was not as important as the quality of the rendered master because the master would then be used to create hundreds, if not thousands, of copies.

The adoption of vector-based objects in high speed electronic printing was governed by:

1. The availability of affordable high speed vector processing chips for use in the printer's RIP.
2. The ability to group vector processing chips to handle parallel object rendering.

Since every page printed is really a *master*, the RIP must be able to render pages much faster, usually 2 – 3 times faster, than the actual speed at which the pages are printed. With inkjet color printers approaching 5,000 (full-color) impressions per minute, the RIP would likely have to render at 40,000 monochrome impressions per minute (5,000 imp x 4 colors x 200%).

Terms Used in Typography

A font generally refers to the physical form of a typographic design or typeface. It can be a case of wood or metal characters or a computer file containing raster or vector information. For our purposes a font is the instantiation of a specific typographic face, height, weight, and style (e.g. Helvetica 12pt Medium Italic). When the first electronic (rasterized) fonts were distributed they were packaged in the traditional way by height, weight and style.

With the introduction of outline fonts there are no longer unique objects for each size and character in a typeface, so today it is common to refer to the entire font family as a font. When licensing fonts be sure to read the fine print to ensure what rights you are granted by the license and what variations are included. Remember that fonts acquired from "free font" sites or compilation CDs may not be licensed for corporate use.

Typeface

Figure 2 - Example of the Frutiger typeface used for signage as originally intended.

A typeface (also known as a font family or even a font) is a series of sizes, weights, and styles that are designed for a common, and complementary, look. A type designer develops a typeface, based on either a commission brief (as in the case of Frutiger) or as a general offering, (as in the case of Avenir). Once the designer has created the typeface it may be distributed through a type foundry or independently. A typeface may include a wide range of variations that include Oblique, Semibold and Ultrabold or they may be limited to Roman, Italic, Normal and Bold.

Font Height

Height is specified in points. Prior to the 1980s the size of a point might have varied by foundry, but since the 1980s when desktop publishing emerged, points have been defined as 72nds of an inch. The firm definition is based on the displays in that era had a 72 dot per inch resolution.

The point size describes the vertical height of the character box in which the character resides. The character box for a font is large enough for the tallest ascender in font as well as the lowest descender. A 10 pt. (point) font would have a character box that is 10/72ths of an inch high.

There are key measurements that are part of describing a character, beyond the point size.

X-height: The x-height is the height of the lower case 'x' in a font. The 'x' is usually the height of the body of all lower case characters, not including the ascender and descender strokes.

Ascender: Ascenders are the strokes that appear above the x-height of the character.

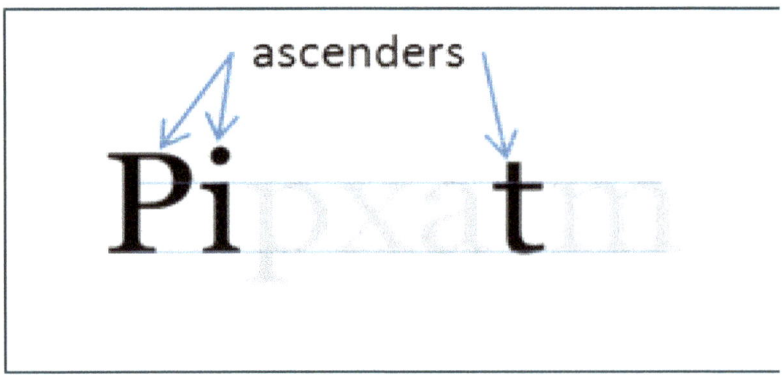

Descender: Descenders are the strokes that appear below the x-height of the character.

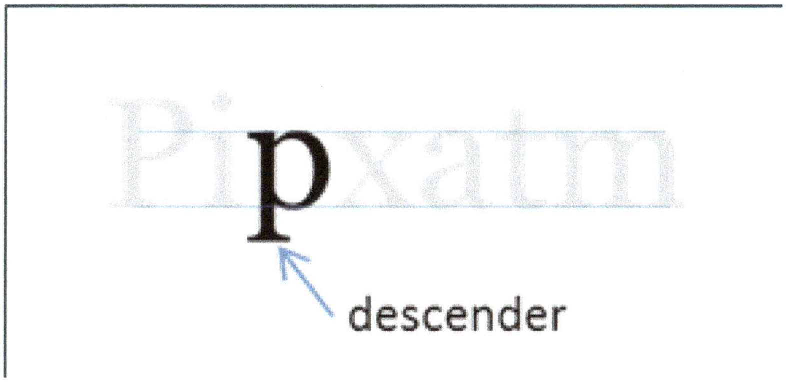

Font Weight

The designer may create additional weights of the font by thickening the strokes within the character. Most fonts will have a bold weight. However, there can many weights within a font, hypothetically up to 800 with OpenType and TrueType. The ISO/IEC 9541 Font Architecture[1] specifies nine weights:

- Ultralight,
- Extralight,
- Light,
- Semilight,
- Medium (normal),
- Semibold,
- Bold,
- Extrabold, and
- Ultrabold.

Pitch

The width of mono-spaced characters, based on the approximate number of characters that can be placed in a linear inch, for example, 10-pitch has approximately 10 characters per inch. Most line-print applications are set for a 10-pitch character set, but there are options for 12-pitch and 15-pitch in some environments.

We say approximate because the mechanical print trains used in true line printers used identically spaced characters. Digital fonts approximate the fixed character widths but there is some variation.

1 ISO/IEC 9541 defines a method of naming glyphs and glyph collections, independent of any document encoding technique; it assumes that one or more methods of associating document encoding techniques with glyph identifiers used in font resources will be provided by text processing systems. ISO/IEC 9541-1:2012 specifies the architecture of a font resource, i.e. the font description, font metrics, glyph description and glyph metrics properties required for font references and the interchange of font resources.

Font Styles

There are three major styles within a font family.

Roman: The Roman, or regular, style is used for text within the body of a document. The characters are upright, not slanted. Most information that we read is set in Roman style.

Italic or Oblique: Characters slope or slant in an italic (or oblique) style. Oblique styled fonts are slated to the right but based on the Roman form of the characters. Italic fonts are also slanted to the right but use different glyphs for the characters.

Condensed: Condensed characters are designed to look similar to the regular face but have a noticeably narrower width. Condensed fonts are often used in signage that could be read at a severe angle, such as shelf labels. We sometimes used a condensed font in a column heading where we want a larger point size that doesn't take as much width.

Ligatures: Designers often include an extended set of characters that are either a combination of 2 – 3 characters (e.g. fl) or a diacritic, a character that is not in our modern alphabet (e.g. æ and œ).

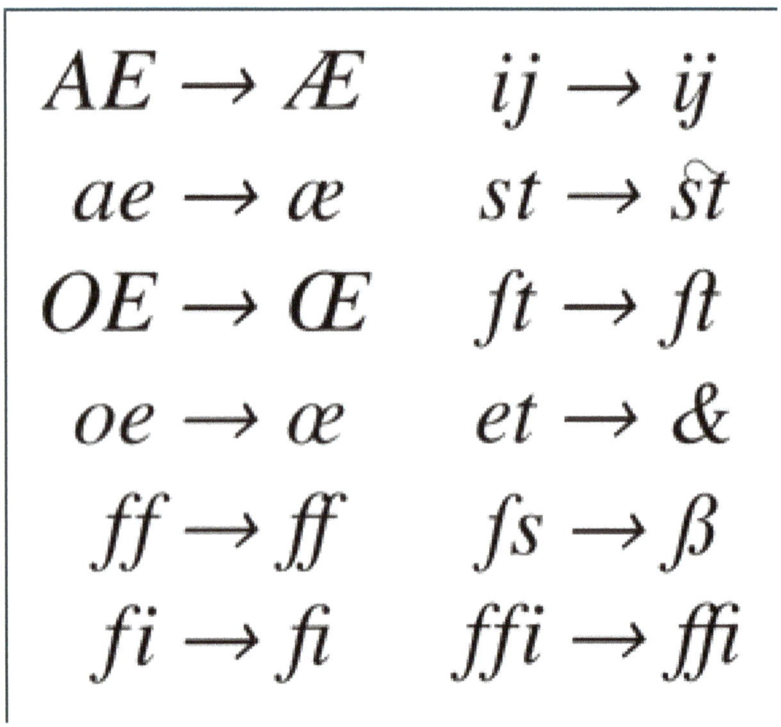

Figure 3 - Ligatures.

The ampersand (&) character was originally a ligature for 'E' and 't' meaning 'et', the Latin word for 'and'. The Trebuchet font's ampersand character, &, is close to the original ligature.

Icons: There are a number of symbol characters that are used to represent words, and often transcend one language. Using icons allows us to express a thought in one character that may take a word, or phrase, in full text. They can be a very effective way to catch the reader's attention and to save space.

Extended characters in most fonts:

- ™ Unregistered trademark
- ® Registered trademark
- © Copyrighted material
- £ UK or Irish Pound
- € Euro

Kerning

Certain pairs of characters look better when they are nudged together. The designer can add 'hinting' to the digital font, describing how to kern a specific pair of characters.

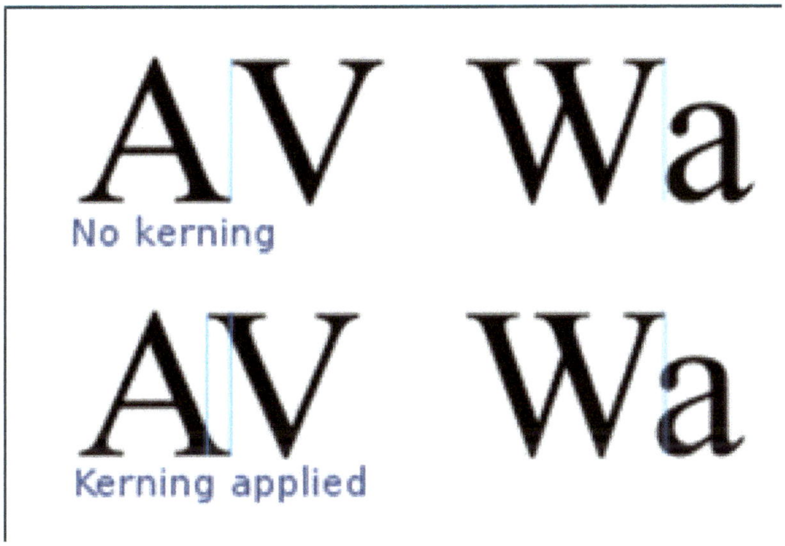

Figure 4 - Kerning.

Typographic Fonts

Typographic fonts are used to present textual information within a document. By choosing the correct font or fonts for a document we improve its readability.

We use fonts for different purposes. With some documents, such as marketing brochures and posters, we use stylized text to capture the customer's attention. With other documents, such as client statements and bills, user manuals, and legal agreements, we want to make the document as readable and comprehensible as possible. Even with the clearest writing we can still make a document unreadable by choosing the wrong font, the wrong size, the wrong spacing, and the wrong column length.

Readability

Contrast: There are a number of things that can make textual information more readable. The most important is the contrast between the characters and the background; you want as much contrast as possible, i.e. 100% black (RGB 000 000 000 / CMYK 0 0 0 100%) on absolute white (RGB 255 255 255 / CIE 145 paper).

Word shape: The next most important is the shape of the words. This is controlled by using mixed case text, providing enough white (quiet) space around the word, and the choice of font face.

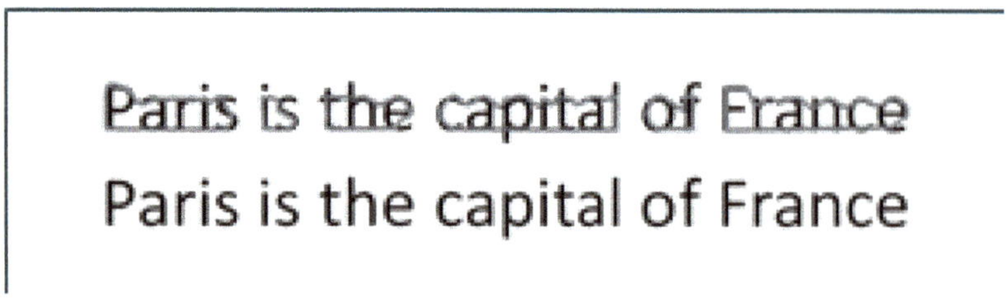

Figure 5 - Word shapes.

As we learned to read we actually were taught to read word-shapes — the series of black strokes within characters positioned close enough together that we recognize a word blob, not individual letters. If there is not enough white space around the words, or the actual shape of the word is unrecognizable because of the shape of the characters, we revert back to looking at individual characters to spell out the word. The most glaring example of this is presenting words in upper case.

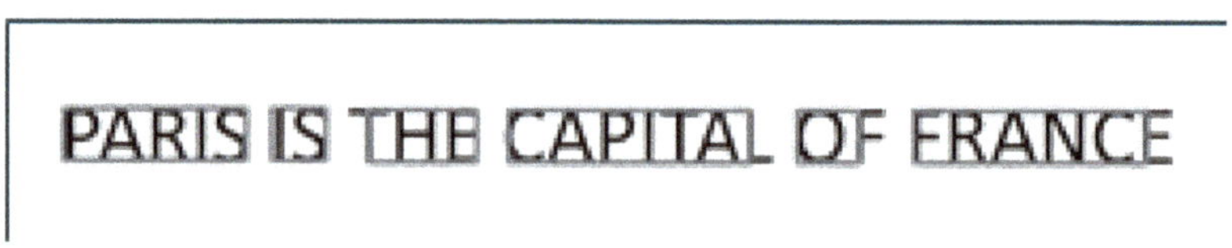

Figure 6 - All caps is hard to read.

Number readability and columnar integrity: Most recipients of transaction documents are most interested in the numeric information presented. It's important that numbers in a font are distinct; they typically are shown as 6 – 8 digits.

Many of the most popular corporate fonts have numbers that are indistinguishable from each other when presented in a seven digit string at 10 pt. The numbers that are easiest to distinguish have 6's with an upward-pointing stroke and 9's with a downward-pointing stroke. Conversely, 6's with strokes that curl back into the character are hard to distinguish from 8's, 3's, and 9's.

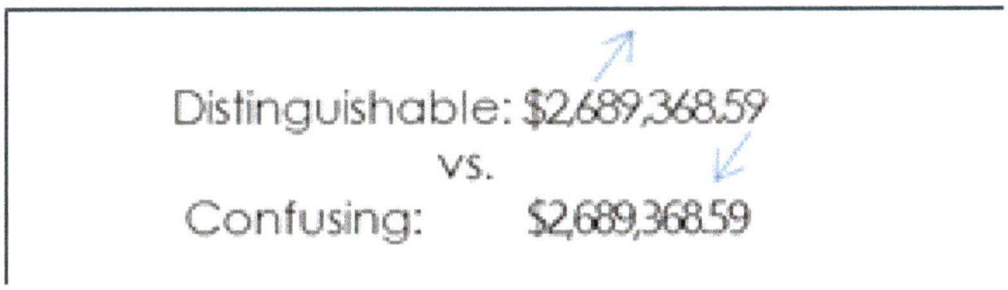

Figure 7 - May your 6's point up.

It is also important that the numbers used in columns all have the same width. Many fonts contain variable width numbers that look good in descriptive text but do not properly line up in a column. Although most state-of-the-art composition systems are capable of mitigating this problem, check that the font contains proper hinting on how to render the numeric character in a column.

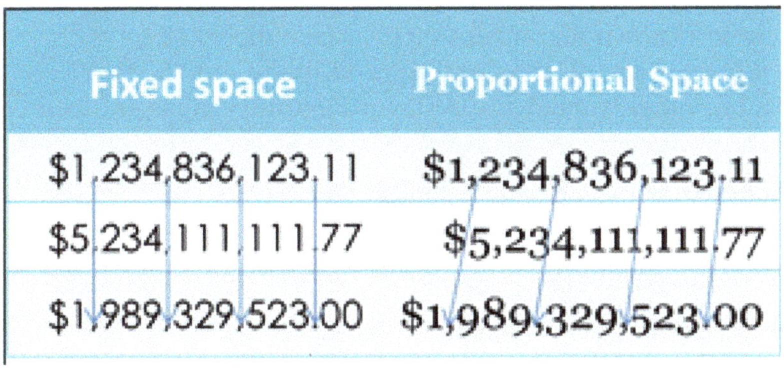

Fixed space	Proportional Space
$1,234,836,123.11	$1,234,836,123.11
$5,234,111,111.77	$5,234,111,111.77
$1,989,329,523.00	$1,989,329,523.00

Figure 8 - Ensure that fixed spaced numbers are used in tables.

Font Structure

Serif vs. Sans Serif: Is it descriptive text, a table or a heading?

Figure 9 - Serifs versus sans serif.

One of the major classifications of fonts is serif versus sans (without) serif. Serifs are the strokes, or glyphs, that are added to the character to provide more differentiation and to speed up the reading process. There are heated debates amongst graphic designers and typographers as to which style is more readable. Many preferences are driven by cultural, geographic, and educational backgrounds.

Serif Fonts

Historically, font faces with serifs were used to present descriptive information (sentences and paragraphs). Most literature is set in a serif font, as are most newspapers.

It's important to understand how our eyes perceive written words. When reading descriptive information, our eyes jump across the line of text, pausing for a fraction of a second to recognize a series of 3 – 5 words, before jumping to next part of the line. This is known as a saccadic jump. The optimal line of textual information is 60 characters (or 8 – 10 words), resulting in 2 – 3 saccadic jumps per line[2]. The eye then travels back through the whitespace below the line to find the next line (a reason to have plenty of white space, or leading, between lines of text).

If the eye does not recognize the word shape(s), the eye moves back on the line to interpret the characters and context (known as regression)[3]. Serifs direct the eye to move quickly across the line in a paragraph of information, and make out the word shapes better. Most adults can more quickly read a 1,000 word article that is set in a serif font[4].

Some of the most commonly used licensed serif fonts include[5] :

- Baskerville
- Bembo
- Bodoni
- Garamond
- Minion
- Palatino
- Caslon
- Century
- Clarendon
- Rockwell
- Sabon
- Times New Roman

Sans Serif Fonts

In the chronology of fonts, sans serif (or groteske) fonts were introduced much later than serif fonts, approximately the same time as the industrial revolution. They were originally intended for signage and posters because they looked cleaner and more industrial. They have become more commonly used for scientific, transportation, and financial tables.

Product and procedural manuals are set in san serif fonts, as is most scientific and engineering research. Through this evolution, sans serif fonts have become associated with quantitative and tabular information.

Some of the most commonly used sans serif fonts include:

- Akzidenz Grotesk
- Avenir
- Frutiger
- Helvetica
- Lucida Sans
- Myriad

2 Typographic Design: Form and Communication 2nd ed. John Wiley & Sons: 1993
3 Typography: how to make it most legible – 5th Edition, pp 16-18, Rolf F. Rehe, Design Research International, 1984
4 The most popular fonts used by designers, Web Designer Depot, Cameron Chapman, 2011. http://www.webdesignerdepot.com/2011/08/the-most-popular-fonts-used-by-designers/Ibid.
5 Ibid.

- Futura
- Gill Sans
- Optima
- Univers

Other Types of Fonts

Script fonts: These fonts (also called cursive fonts) mimic calligraphy and handwriting. They are used to mimic handwritten notes or a signature. They should be used sparingly.

Ornamental fonts: These fonts contain special characters that are not found within a standard font. These characters, like a finger pointing, can be used in bulleted lists to embellish a block of text. They are not usually used within unbulleted descriptive (business) information. Common ornamental fonts include dingbats, web bats, and ornamentals. Examples of their charactersets are shown below:

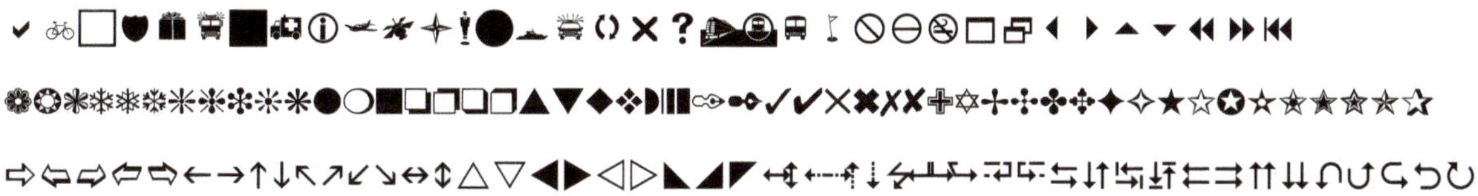

Ornamental fonts should be used sparingly because they add an additional series of objects to manage and render at the printer. They can also impact search indexing as the hex value for the character may be the same as an alphanumeric character in a regular font.

Asian Language Fonts

Written Asian languages, including Chinese, Japanese, Korean, and Vietnamese, each have thousands of characters. Traditional Latin language fonts contain less than 255 characters, which allows them to be identified in a single byte. Asian language fonts use multiple bytes to identify a character, allowing hundreds of thousands of characters within the font . The term double-byte is used to describe fonts that use 2 bytes per character.

Multi-byte fonts are usually issued for each Asian language, Hànzì in Chinese, kanji in Japanese, hanja in Korean, and Chữ Nôm in Vietnamese. However, you can also buy Chinese / Japanese / Korean (CJK) packs. The Asian language fonts include all major Latin, Cyrillic, math, and financial characters.

Asian languages can be written either horizontally or vertically. Asian characters sit in a fixed square space (including the Latin characters). Most business correspondence is written horizontally, but there could be instances where an informational, educational or marketing message could be written vertically.

Although most European and North/South Americans may not think about the inclusion of Asian fonts in a document production system, there is a growing need (especially within Wealth Management) to provide client support messages for Asian customers in their language of choice. Also, Asian multinationals have developed worldwide client-support systems that use full Asian language (double or multi-byte) databases, and generate transaction documents that use Asian language fonts.

There are many different formats for Asian language fonts:

- multi-byte,
- singletons, lead unit, trail units,
- double-byte,
- IBM DBCS (EBCDIC based),
- shift in/shift out,
- CJK DBCS (ASCII based),
- Big5, GB, S-JIS, KSC,
- Unicode,
- UTF-8 = variable-width encoding (maximizes compatibility with ASCII),
- UTF-16 = variable-width encoding, and
- UTF-32 = fixed-width encoding.

Choosing the Correct Font

We choose the fonts based on a number of factors.

Style Guidelines

Corporations and governments often have preferred fonts that they use in all customer communication. These fonts are often specified in their corporate style guide, which is usually available from the corporate communications department. Company brochures, signage, and advertising text will typically use a single licensed font[6] (or one serif font and one sans serif font). Because web font formats are not always available the brand guidelines may also call out a web font that has a similar look.

Whenever possible, the mail owner should use the same typographic guidelines for transaction documents as they would use in other printed customer documents.

Most current composition systems and production printers can properly exploit current vector-based licensed fonts. However, many legacy applications that create line-print files, Xerox Metacode files, or very early versions of AFP files may require a legacy raster font to print. While some have similar names to the popular fonts like Helvetica and Times, they are only approximations.

Most licensed fonts have not been easy to use in a web environment. As a result, document owners will sometimes choose one of the core fonts within HTML. With the more recent levels of HTML (4 and higher), some licensed fonts can now be incorporated into webpages. However, the legal issues associated with web font licenses may still be an inhibitor.

One of the advantages of presenting electronic documents in PDF format is that licensed fonts (and font licensing) can be easily handled.

6 Font designers usually copyright their font designs, trademark the name, and then license document producers to use the fonts under specific conditions. For example the Frutiger font was designed by Adrian Frutiger under commission to the Charles de Galle Airport. The font is distributed primarily by Linotype (now owned by the parent of Monotype). The licensing agreements for Frutiger (and other Monotype licensed fonts) can be found at www.fonts.com.

Font Use

Although there are no strict rules on serif versus sans serif usage, there some conventions, based on consumer familiarity:

- headings are typically in a sans serif font,
- tabular information is usually in a sans serif font,
- telephone numbers, email addresses, street addresses, and websites are usually in a sans serif font,
- descriptive information is usually in a serif font, and
- descriptive call-outs are usually in an italicized serif font.

MICR

MICR is not a font, but a technology with several components. The Magnetic Ink Character Recognition (MICR) process uses magnetically charged characters that are read at extremely high speeds. The read-head on the sorter is similar to the read-head on a magnetic hard-drive; it recognizes variances in the magnetic properties of the characters that it's reading. Unlike an Optical Character Reader, it can misread perfectly formed characters if the ink (or toner) does not have the correct magnetic properties. Conversely, it can read malformed characters, as long as the magnetic properties of the characters are correct.

There are two typefaces used to create characters with the right magnetic properties. In the United States of America, Canada, Australia, United Kingdom, Japan, India, Mexico, Colombia, and Turkey, the E13B font (promoted by IBM) is used to encode checks and other payment instruments (often referred to as 'presentments'). The CMC-7 font (promoted by Groupe Bull) is used in France, Spain, Israel, South America (except Colombia) and other Mediterranean Countries.

MICR printing remains machine readable, even when over-stamped, marked, or mutilated.

MICR is most commonly used for checks, however, many checks are captured as images as they enter the bank clearance systems, and the images are then processed as though they are physical checks. MICR is also found in direct mail applications.

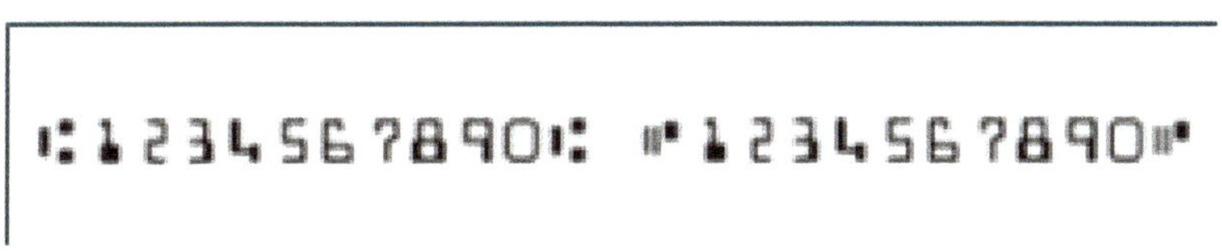

Figure 10 - e13b font. One of two used for MICR sorting.

Common Font Object Formats

The following are common font object formats. They are listed in reverse chronological order:

OpenType

Most licensed fonts are provided in the OpenType font format. The format was developed by Microsoft with Adobe's participation, and announced in 1996. OpenType exploits vector commands (commonly referred to as outline data in the font world). The creators wanted the structure to take advantage of more advanced 'quadratic' vector commands to provide finer detail within a character, and better handle the complexities of non-Latin languages. "OpenType fonts can also include typographic refinements such as true small caps, different styles of figures, and extensive sets of ligatures and alternates, as well as complete sets of accented characters and diacritical marks.[7]"

OpenType was intended as a replacement for TrueType and Adobe font formats. To do this, the predecessor fonts needed to be supported. Also, it has evolved to provide extensive web and tablet support. As a result, there are a number of sub-types of OpenType:

- **Open Font Format (OFF):** ISO adopted a version of the OpenType format in 2007[8] known as Open Font Format (OFF). Unless otherwise specified, an OpenType font will be OFF formatted.
- **Web Open Font Format (WOFF):** WOFF is a font format that is used within webpages by most browsers. It became a W3C Recommendation in late 2012[9]. The format essentially wraps TrueType, OpenType and Open Font Formats with metadata that allows it to be downloaded, used and erased after use. The fonts are compressed within the transmitted object. The WOFF process enforces a same-origin policy to ensure that a downloaded font does not be used by other (improperly licensed) websites, or by applications on the recipient's computer.

OpenType supports outline fonts that were originally created for TrueType or for PostScript (CFF). The formats are:

- **ClearType rendering technology:** Microsoft introduced the ClearType subpixel rendering technology to improve the edges of characters on high resolution displays. It's available on all Windows operating systems after XP. The technology can render OpenType and TrueType fonts, and there are a limited number of fully defined ClearType fonts available.
- **TrueType:** The TrueType[10] format was developed by Apple and Microsoft in the late 1980s in competition to Adobe's PostScript (Type 1) font format. TrueType provided a higher level of scalability for outline fonts compared to Type 1.

The major technical difference is that TrueType uses quadratic Bezier curves, which seem to render quicker, with a higher level of scalability, than cubic curves.

Adobe PostScript Fonts: The Adobe PostScript font was designed by Adobe for professional typesetting. The fonts were described in vector commands called cubic Bezier curves, originally used with graphic plotters and radar screens. These cubic commands do not easily render into a raster environment (common displays and laser printers).

- *Type 1:* The initial format for Adobe fonts was called Type 1, and was intended for graphic arts applications. The rendering process can be quite slow due to the number of commands required to create a character using the cubic structure.

7 What is OpenType: http://www.microsoft.com/typography/WhatIsOpenType.mspx
8 ISO/IEC 14496-22 (MPEG-4 Part 22)
9 http://www.w3.org/TR/2012/REC-WOFF-20121213/
10 True Type Specification 1.66 http://www.microsoft.com/typography/SpecificationsOverview.mspx

- *Compact Font Format (CFF) - Type 2:* In order to improve the speed of rendering to screen and to electronic (lower resolution) printers, Adobe introduced CFF, in conjunction with Adobe Acrobat (Version 3) and Portable Document Format v1.2 (PDF). The CFF format still uses a 'limited' cubic structure that subsets to only the necessary commands for 'low resolution' rendering. This speeds up the rendering process without visual degradation.

AFP Font Object Content Architecture (FOCA): Most transaction documents are produced using Advanced Function Presentation (AFP) file structures. High speed production printer RIPs used the AFP object structure to optimize rendering speeds and fidelity.

- FOCA is the object type used to AFP-specific fonts within the AFP environment[11] . These fonts have been specifically designed for AFP presentation environments. They include:
- *38PP fonts:* The 38PP fonts were the fonts designed for the initial implementation of AFP in the mid-1980s.
- Other fonts supported in the AFP architecture: The AFP architecture (but not FOCA) also supports other font-types as data resources, including:
  - OpenType
  - TrueType

Core fonts: Microsoft includes a series of fonts within the Windows operating system. The W3C also specifies a series of fonts that are to be included with any W3C HTML (level x and higher) server. Both are known as the core fonts. Personal printer manufacturers usually include a core font set on their printers, based on the then current Windows core fonts.

Many applications will assume that a core font is present on the printer unless you otherwise specify. Problems can occur if Windows removes the font as part of a new version; or if you change the targeted printer, and it does not have the font.

Windows: Microsoft Windows provide a set of fonts to be used with Windows applications[12]. Microsoft updates the core font set with each major Version. You can check the Microsoft font page to find out which fonts are provided with each operating system.

Web fonts: Microsoft led a project to introduce a family of core web fonts. Although the project was cancelled in 2002, many HTML servers still support references to the following fonts. With the implementation of WOFF, the requirement for predefined fonts is disappearing:

- Andale Mono (replaced by Lucida Console on later Windows operating systems)
- Arial
- Arial Black
- Comic Sans MS
- Courier New

11 Font Object Content Architecture Reference, 2005 http://afpcinc.org/site/assets/files/1129/ibm_foca_focaref5.pdf
12 http://www.microsoft.com/typography/fonts/

- Georgia
- Impact
- Times New Roman
- Trebuchet MS
- Verdana
- Webdings

Input: Where do fonts come from?

Most fonts are created by professional font designers who sell (via a user license):

1. Through distributors (often called foundries, an artifact from the days when font characters were made out of steel). Major font distributors include: Adobe, Monotype (including Linotype and Fonts.com subsidiaries).
2. As part of another software product, e.g. Microsoft Windows, Microsoft Office.
3. As part of a print management system, or imbedded within a printer's firmware.

Output: Where do we place fonts for use in the production process?

Font objects are loaded into production libraries associated with:

- composition system software,
- print stream utility software, and
- print management systems.

The software application imbeds the font object into the print stream prior to the print stream being presented to the RIP process for rendering.

Graphics

Graphics and Comprehension

We use graphical representation to improve document comprehension, and as a complement to textual information.

- Graphical branding, such as corporate logos.
- Business graphics, to illustrate critical information:
  - Pie or Bar charts are used to show distributions,
  - Line charts are used to show activities over time.

Vector commands have a number of advantages over raster graphics, including faster handling by Raster Image Processors (RIPs) to render logos and business graphs. Vector graphics also produce cleaner edges when scaled.

Graphic Object Types

There are hard and soft graphic objects. Hard objects are developed and pre-tested as discrete objects, such as corporate and product logos. The object is stored in final format and then loaded into the print stream as a defined resource that can be rendered once and then repetitively placed.

Soft business graphics are generated during composition based on stored business rules. The RIP renders the graphic and places it immediately.

The most common graphic types are:

- **Scalable Vector Graphics (SVG):** SVG is a World Wide Web Consortium (W3C) recommendation that provides two-dimensional vector graphic support[1]. It allows objects to be described using graphic commands. All current web browsers support SVG and most smartphones and tablets support the SVG mobile subset. SVG can include pre-rasterized images and vector fonts. Not all composition systems, Digital Front Ends on printers, or RIPs support the SVG object type.
- **Encapsulated PostScript (EPS):** Most graphic arts applications support EPS as an output format. Although EPS files are usually formatted pages or documents, you can use it to develop objects, including logos. You can include vector commands, pre-rasterized images and vector-based fonts.
- **AFP Graphic Object Content Architecture (GOCA):** GOCA is the AFP object type that supports generic vector-graphics, such as business graphs, forms, and templates (also called overlays)[2]. There is also a specialized object type for barcodes (called BCOCA) that uses most of the commands within this format. The GOCA format uses draw rules, such as Line, Full Arc, Partial Arc, Fillet and Cubic Bezier Curve.

Input

Most graphic design tools support graphic object output formats. These include Microsoft PowerPoint (PPT native format, XPS, PDF, and XML) through to Adobe InDesign (EPS, PDF). Hard objects (saved in a supported format) can be created using a design tool and then promoted into a production environment. Check your composition tool and/or printer RIP to see what graphic object types are supported.

Composition and/or print stream utilities can generate unique soft objects that are imbedded within the print stream. GOCA objects will be imbedded in AFP files, and EPS could be imbedded in PostScript.

Output

Hard objects are pre-designed, tested and promoted to production resource libraries. Soft (unique to document) objects are generated with the creation (or modification) of the document and carried within document.

1 W3C Scalable Vector Graphics (SVG) 1.1 Specification http://www.w3.org/TR/SVG/ . Version 1 was adopted in 2001 and Version 1.1 in 2003. SVG 2 is scheduled for recommendation in August 2014, and will better integrate with CSS, HTML5 and WOFF. It is scheduled as "recommendation" for August 2014, according to the W3C SVG Roadmap.

2 Graphics Object Content Architecture for Advanced Function Presentation Reference, January 2012 http://afpcinc.org/site/assets/files/1130/goca2ref.pdf

Barcodes

We use barcodes on documents, packaging and signage to provide machine readable information. This can be done to:

- assist in the tracking of the document though the manufacturing and distribution process,
- identify the document to the sender, the recipient or interested third parties (IMB codes in address block),
- create a call to action as a part of a marketing campaign (QR Code on a brochure or advertisement),
- carry information in machine readable format that is also presented as text (QR Code with business card information), and
- allow downstream applications to validate the authenticity of the document or capture information off of it.

Reading Barcodes

Barcodes can be read with either an LED Scanner or a digital camera.

LED Scanners fire a constant light signal at the barcode area, and then read the returning light signal that reflects off of the white background (black ink or toner will absorb almost all of the transmitted light). The reader mechanism converts the light/no light signal into on/off bits. It then translates the on/off bits into alpha-numeric information.

Cameras take a picture of the barcode and then convert the binary matrix into alpha-numeric characters.

Check Bars

Most barcodes use specific bars for registration and validation. Linear barcodes use start and end bars. 2D barcodes use track points, QR codes use three corner squares called Position Detection Patterns, and Data Matrix codes are delineated by two solid adjacent borders in an "L" shape (called the "finder pattern") and two other borders consisting of alternating dark and light "cells" or modules (called the "timing pattern").

Barcode Topography and Readability

Every barcode schema has minimum specifications to maximize readability. The best standard to follow is the one provided in the manufacturer's specifications for the reader being used. If you don't know the reader, you should use the industry specification for that barcode. Remember that the faster the reading process, the tighter the tolerance to the spec.

Specifications include:

- **Orientation:** The direction in which the barcodes can be read. Most linear barcodes can be read in the 0 and 90 degree orientation.
- **Minimum height:** The minimum height of the longest stroke. This will be expressed in inches or millimeters (mm) in the barcode reader specification,

although some composition systems may express it in point size (72nds of an inch).

    - Some barcodes (e.g. 4SB) have strokes with ascenders and/or descenders. The minimum height would be the height of the stroke with both.

- **Minimum stroke width:** How wide the black strokes must be, and how much is required between them, to be read. This is usually expressed in inches or millimeters.

- **Minimum contrast:** The minimum percentage difference between the light (white[1]) background and the dark (black) background. Although most barcode readers work best with 100% contrast, there are tolerances for off-color paper or non-black ink.

    - In addition, certain paper characteristics can degrade readability:

        - shine caused by calendering or chemicals,
        - brightness[2],
        - rough finish, and
        - surface warping during printing.

- **Minimum quiet zone:** There needs to be minimum background space round each barcode that is free of anything that contrasts, such as variable print or preprinted information or decoration.

  You should also ensure that the reverse side is also clear of anything that contrasts. Almost all paper will have some opacity under LED light signals.

- **Vector-object barcodes:** Barcode readers require good dark to white transition that can be best achieved with crisp line edges. Printer RIPs (and video displays) can best render 'clean' barcodes that are described in vector commands.

Barcode Formats

Character Barcodes

The first barcodes were character-based. This was because impact line-printers and early laser printers could only print a series of pre-rasterized characters that had the vertical or horizontal stroke (or strokes) across the entire character box. If identical characters were printed on top of one another you could create a larger barcode. The problem with a character-based barcode is that the RIP has no idea that it is a barcode and cannot properly tune the strokes. Barcode fonts should be avoided.

Image Barcodes

As barcodes became more complex, software was designed to render barcode images that could be presented to the RIP as a binary image. These utilities were good for UPC and other complex interleaved barcodes. They were designed to work on 300 dpi devices with consistent optical density. As printers ran faster and increased to 600 dpi, the barcode image generators had to be updated for the new technology. Problems could occur between vendor technologies (different RIP algorithms; different optical density) because the RIP would likely use the binary image as is. Scaling or rotating the barcode

1 Absolute white paper should measure 154 on the CIE Color Gamut.
2 Most xerographic paper has 'blue' fluorescent whitener added to make it 'brighter'. As a result 'brightened' paper can actually measure above 100% reflectivity due to the introduction fluorescent blue dye.

image can completely change its printing characteristics. Barcode images should be avoided.

Vector Graphic Barcodes

Most production printer RIPs can render cleaner and more faithful barcodes from vector commands. Some of the presentation architectures actually support barcodes as a native object-type. Others use the SVG format.

AFP BarCode Object Content Architecture (BCOCA)

The most common native barcode format used for Transaction Document Production is BCOCA. With the object architecture, barcode objects need only to provide the following information so that the RIP can render it:

- schema,
- box size, including quiet space,
- orientation,
- X/Y position on the page, and
- information to be encoded.

Barcode Schema

There are many barcode schemas with different functions and purposes. Some have been developed by specific industries (e.g. UPC by retailers, 4SB by the International Postal Union) or by standards groups (EAN-13 by GS1).

Optical Mark Recognition (OMR)

OMR is the first and most primitive form of variably encoded barcodes. It was implemented prior to the availability of laser printers and deploys two standard characters, a pair of horizontal underscore '_' [3] for barcodes that run vertical and a single vertical line '|'[4] for barcodes that run horizontal.

This barcode is still used with inserting systems on older automated mailing applications. It has limited functionality for the amount of page real estate that is uses, compared to newer barcode schema.

3 ASCII x5F and also called the 'low line'; not the 'em-dash' x97 nor the 'macron' xAF
4 ASCII x7C

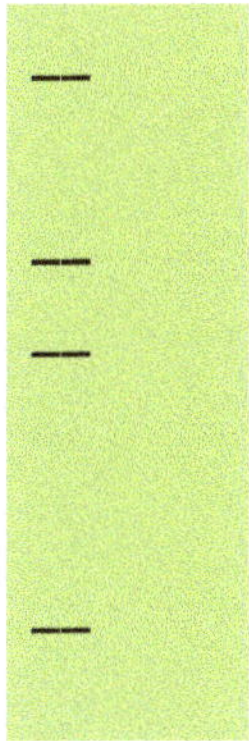

Figure 11 - Vertical OMR layout with underscore characters shown in a shaded block. The shading is not relevant.

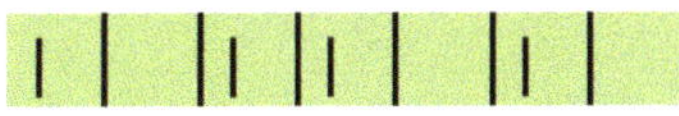

Figure 12 - Horizontal OMR layout with vertical line characters.

Linear Barcodes

Interleaved 2 0f 5

0097809789

Figure 13 - Interleaved 2of 5 barcode.

Code 39 (3 of 9)

Code 39 barcode is to track documents through the manufacturing process; primarily at the inserting stations. The schema was developed for Boeing by Intermec to track airplane parts.[5] It was standardized by American National Standards Institute (ANSI) in 1983.[6]

It is built from 39 characters plus additional control characters consisting of the uppercase letters A-Z, the numbers 0-9; space, and special characters: $ / + %. It does not contain a check digit which means that it is more susceptible misreads. It also takes more space to present information than later versions (see below) or 2D barcodes.

The US government specification, LOGMARS[7], is based upon Code 39.

5 Introduced in 1974 as a font.
6 ANSI MH 10.8 M-1983
7 ANSI/AIM BC1/1995, Uniform Symbology Specification - Code 39

ABC123

Figure 14 - ANSI Code 39 barcode.

Code 93

This is a compact version of Code 39 that is not widely used.

ABC123

Figure 15 - Code 93 barcode.

Code 128

The Code 128 barcode provides higher density information than Code 39. It uses the 128 ASCII character set, and can use the full Latin-1 character set. The GS1-128[8] barcode subset is used worldwide for product container and palette tracking.

Abc123

Figure 16 - Example of a Code 128 barcode.

Universal Product Code (UPC)

There are a number of variations of UPC. They are used to identify products for sale within retail stores. The barcode is mainly used at the point of sale to add its price to the sales transaction.[9]

UPC-A

This barcode used to identify standard products within retail stores. UPC-A consists of a unique 12 digit number, based on the Global Trade Item Numbers (GTIN) specification. The barcode uses 13 characters plus a check digit. There is room in the specification for 2 or 5 optional digits.

8 GS1-128 barcode specification http://www.gs1.org/docs/gsmp/barcodes/GS1_General_Specifications.pdf
9 note: and the receipt presented is an actual 'transactional document'.

It was developed by the US Uniform Grocery Product Code Council[10], and first implemented in 1974. They are widely used in used in the United States, Canada, the United Kingdom, Australia, New Zealand, and in other countries.

0 21000 75896 8

Figure 17 - UPC-A barcode.

EAN-13

The European Article Numbering (EAN) 13 barcode is a superset of the US developed UPC-A barcode. The barcode adds an extra digit at the beginning of the code, allowing for 10 trillion variations.

The EAN-13 code can be used with most North American UPC scanning systems.

9 780978 945619

Figure 18 - EAN-13 barcode..

UPC-E (End-bit)

The UPC-E barcode is used to on small packaging on which an 'in spec' UPC-A barcode will not fit. It cannot be used as a substitute for EAN-13. It uses the end-bit to convert the binary value of the displayed six digits.

0 210007 4

Figure 19 - UPC-E barcode.

10 Now known as GS-1 US http://www.gs1us.org/

Two Dimensional (2-D)

PDF 417

The PDF-417 font is used by a number of industries and government organizations. The airline industry uses it for boarding passes. The US Department of Homeland Security has selected it for RealD[11] compliant drivers' licenses and other identity cards. It was developed by Symbol Technologies in 1991 and is an ISO standard[12]. It supports all ASCII characters and includes error correction up to about 1850 ASCII or 2725 numeric characters.

Figure 20 - PDF-417.

Data Matrix

The Data Matrix barcode is used for labeling small items (e.g. electronics) or items where the barcode is unobtrusive (e.g. documents within a manufacturing process). The US Electronic Industries Association endorses its use for electronic parts identification. It is also used extensively within the food industry It can carry a great deal of information in a small space.

The barcode is an ISO standard[13]. It uses dark and light cells, with the left and bottom edge cells all darkened (to indicate the orientation). The information to be encoded can be up to 2,335 characters of alphanumeric data, including error correction codes.

Figure 21 - Data Matrix barcode.

11 US Federal REAL ID Act of 2005. http://www.gpo.gov/fdsys/pkg/PLAW-109publ13/html/PLAW-109publ13.htm

12 ISO standard 15438

13 ISO/IEC 16022:2006—Data Matrix bar code symbology specification ISO/IEC 15415—2-D Print Quality Standard

Aztec (also known as Bulls-eye)

Figure 22 - Aztec barcode.

Quick Response (QR Code)

Figure 23 - QR Code barcode.

3D

High Capacity Color Barcode

Figure 24 - High Capacity Color barcode.

Disguised Barcodes

Microglyph (formerly Dataglyph)

The microglyph format can be used to digitally sign documents in a way that is unobtrusive. A microglyph object could be applied, as a hard resource, to a series of documents to authenticate the issuer of the document. A unique microglyph could be generated for each document, as a soft resource, to authenticate both issuer and the intended recipient.

The Dataglyph was invented at Xerox Palo Alto Research Centre, and offered by Xerox as a proprietary process[14]. Xerox has since licensed Microglyph Technology GmBH[15], who enhanced the technology and renamed it 'microglyph®'.

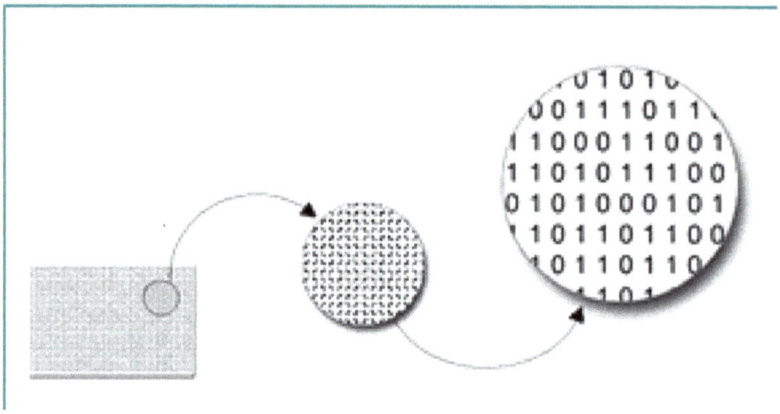

The schema uses extremely small forward '/' and back '\' strokes of different widths to simulate pixels within a photo or line art. The barcode interpreter determines which way the stroke slants, and translates them to bits. The bit stream is then converted to data. The data is usually encrypted.

Raster Objects

One of the easiest ways to describe information to print is as an image. We have been transmitting and printing documents since the early 1920s. The first technology, Wirephoto, used an analog signal to scan and transmit black on white images to the recipient device. These technologies, along with analog television cameras, were the starting-point for image capture technologies and the standards that morphed into many of the digital raster image object-types that we use today.

In the early days of laser printing we used these technologies and formats to create most of the document objects we needed. Raster objects minimized rendering, but rasterized fonts, logos and business graphics were pre-rendered to the resolution of the printing device and potentially tuned to its idiosyncrasies.

If a print job was targeted to a print device with a different print resolution the rasterized objects had to re-rendered in the new resolution. A common challenge for companies using AFP workflows came when the common machine resolution changed from 240 dots per inch squared (dpi) to 300 dpi. Re-rasterizing of any 240 dpi fonts, logos or raster graphics was a time-consuming process fraught with potential errors.

14 Can only be rendered on Xerox RIPs
15 http://www.microglyphs.com/english/html/company.shtml

Attempts to transform from one resolution to another often produced mediocre results. Rescanning was not always an option as original artwork was often not available, and many object issues were only discovered during production testing (or even in actual production).

Today, we try to limit the use of raster objects to scanned documents and signatures (usually stored as TIFFs), and pictures (stored as JPEGs).

The best practices is to avoid using raster objects for fonts, barcodes and business graphics or line drawings and use vector-based objects. Printer RIPs can render vector versions of these objects more accurately than raster versions.

Scanned Raster Objects

Organizations capture an enormous number of paper based documents via scanners. Scanned documents can be more easily managed as part of business workflow and/or archived as corporate records. These documents are usually stored in either TIFF or PDF Image[16] format.

Outbound documents often include scanned images from these systems. Examples include check images printed as part of a monthly bank statement, client-signed shipping manifests printed as backup to an invoice, and insurance applications appended to the issued policy.

To effectively use these scanned pages as document objects, it is important to understand in what format, and at what resolution, the original document was stored. Many documents are scanned at a low resolution (readable on a display, but fuzzy when printed at 600 DPI) to minimize storage and transmission size. They will look like a FAX if printed at their original size since most RIPs will not apply anti-aliasing to TIFF (or JPEG) objects under the assumption that this could be interpreted as tampering with the original rendition.

Tagged Image File Format (TIFF)[17]

TIFF was originally developed by Aldus Corporation, which was acquired by Adobe Corporation in September 1994. It was promoted as a universal scanning format for desktop publishing and was adopted by most scanning technology companies in the mid-1980s.

Its original format was binary dot placements (1 for black, 0 for white), but evolved to support grey-scale dot placement (using either a half or full byte to describe the shade of grey)[18], and eventually color.

It is the preferred format for storing scanned documents because it uses lossless compression algorithms as its default. You can use the JPEG (Lossy) compression format within TIFF when encapsulating photographs.

There are many extended versions of TIFF generated by vendor software. A best practice is to use Baseline TIFF[19] as the RIP may not be able to faithfully render extended functions such as layers or JPEG and LZW compression.

16 The document can be stored as a multi-page TIFF Each page is a separate raster image.
17 For additional TIFF material consult: http://partners.adobe.com/public/developer/tiff/index.html#spec
18 A half-byte describes 16 and a full-byte describes 256 shades.
19 TIFF Revision G Final http://partners.adobe.com/public/developer/en/tiff/TIFF6.pdf

TIFF Extensions

TIFFs can have multiple sub-files (usually pages of a document) within the object. Multi-page TIFFs require pre-processing to generate individual object for each page before use as document objects.

The most popular compression formats for scanned documents are CCITT G3.1 (specified within base-line TIFF), and CCITT T.4 / T.6 (allowed as supplier extensions, but not universally supported). It is important known whether the page images to be used are the latter and if this impacts the RIPs rendering of the object.

You may run into TIFF files that are compressed using Apple's Packbits format (specified within baseline TIFF).There are also TIFF extensions that support uncompressed CMYK layers (4 layers, each supporting an 8 bit color channel) within an object. Vendors can also add private tag metadata to the TIFF file to exploit proprietary functions.

For any extensions to TIFF it will be important to determine that the RIP will be able to process it without error.

Scanned Portable Document Format Files

Adobe controls both the PDF and TIFF format. The TIFF format has been stabilized since 1996, and that format is used as a subcomponent within PDF files.

Most new document scanning applications use the PDF format as the container for a scanned document. Within that document, PDF uses Baseline TIFF images for each scanned page. The scanned document can then contain metadata such as indexing information, and text information that emerges from Optical Character Recognition.

As with multi-page TIFFs, you will need to pre-process these PDFs to get to the individual TIFF objects. Most likely you'll need to deconstruct the PDF files into a series of TIFF objects. Within the AFP environment, however, you can pass the entire PDF file to the RIP and instruct it to render only on a specific page. This avoids the need to deconstruct PDF image files.

Joint Photography Experts Group (JPEG) Format

JPEG is both the common lossy compression format for digital photography (used within many different image file formats, such as TIFF and RAW), and a file format in its own right. The latest JPEG standards can be found in ISO/IEC 10918.[20]

The JPEG format accurately renders photographs and paintings with realistic variations in hue and color. However, it is not optimized for line drawings or scanned documents.

JPEG photographs captured using digital cameras are saved in the JPEG Exif format, which is restricted to 24-bit color (8 bit shading for the R, G and B color channels). However, many professional cameras now support 32-bit color (8 bit shading for the R, G, B and Alpha (pure white) color channels). 32-bit color photos may have the .jpg file extension but are likely RAW format inside.

20 The latest section is ISO/IEC 10918-6:2013 -- Digital compression and coding of continuous-tone still images: Application to printing systems.

Exif formatted photos do not contain an ICC color profile. You will need to add the desired color profile to the JPEG metadata before using it as an object.

There is a wide compression range within the JPEG format. The highest quality uses all 8 bits for each color dot placement (plus about 10% overhead, making the file size about 9 bits per color dot). Lower quality photos can be scaled down to less than 1 bit per color dot. Photos with severe compression will not render well on color printers (although they will look fine when viewed on a smartphone or tablet).

It is important to understand that the following items will impact speed and fidelity of rendering at the RIP:

- The source picture size is the same as the size when placed. Although RIPs can rescale JPEGs, this takes an enormous amount of computing, and the results can be disappointing. If you want to place a 1 inch square picture, make sure it's pre-rendered to that size.
- The severity of compression. The less the compression, the more faithful the rendition and the less likely that you'll see artifacts on color boundaries.
- 32 bit color (RGBA). Although RGBA photos look more colorful when viewed on the latest LED displays (and will look amazing on the soon-to-be-released OLED displays[21]), the Alpha channel can cause havoc at the RIP.

RAW Image Format

These files (sometimes called digital negatives) are the minimally processed photographic files captured by a camera or scanner. RAW files usually have a much wider color gamut than JPEGs. However, there are numerous RAW formats used amongst the camera vendors. For that reason, they must be further processed to be usable, usually transformed into (or encapsulated within) JPEGs.

AFP Image Object Content Architecture

"The Image Object Content Architecture (IOCA) has been formulated to provide a format suited for high speed printing. IOCA contains enough flexibility that a wide variety of images can be printed, but formats images in such a way that they can be printed efficiently and with minimal processing."[22]

IOCA provides a common container for the many different types of image formats, allowing them to be properly rendered by the RIP.

IOCA supports AFP-specific as well as many of the formats described above. However, TIFF, JPEG and other non-AFP file types must be encapsulated in an IOCA wrapper to be used. Image types supported include:

- JPEG algorithms,
- TIFF PackBits,
- TIFF LZW,
- CCITT Group 3,

21 OLED display technology uses four LED colors for each pixel, including pure white. This allows it to faithfully replicate the Alpha channel in RGBA photography (and video).

22 Image Object Content Architecture Reference - Release 6.0 - S550-1142-00, page 7. http://afpcinc.org/site/assets/files/1128/ibm_ioca_55011420.pdf

- Adaptive Bi-level Image Compression (ABIC) – extensively used in bank check image capture systems
- InfoPrint MMR—Modified Modified Read
- No compression
- Run Length 4
- G3 MR—Modified READ
- G4 MMR—Modified Modified READ
- JBIG2

Color is extensively supported in IOCA Function Set 45.[23]

Page Segments (PSEG)

The original format for (uncompressed) IBM AFP image objects.

.IMG

The format for image objects with the Xerox Metacode environment.

Other File Formats

You may run across images in the following formats. Most likely you will need to transform them into TIFF or JPEG format they can be rendered.

- GIF
- BMP
- PNG
- PPM, PGM, PBM, PNM and PFM
- PAM

Input

Image objects are usually captured using

- Digital cameras,
- Scanners, and
- Graphic design software.

23 Functional Set 45 describes bilevel or color tiled images.

Compound Objects

The documents we build today are like Russian matryoshka dolls; we nest logical subset objects within larger objects. For example, within a file containing thousands of documents, we find each document nested inside with a beginning and ending tag. Within each document, we find each page with beginning and ending tags. Within a page we find text block, graphic and image objects presented (or referenced).

The advantage of building documents out of nested objects is that it is easier to develop each object independently. We can build the document's address block object independently and then place it into position independent of other page objects (such as tables, messages, contact information, logo, page number, and packaging barcode) to compose the page. This allows programmers to change the position of the address block without having to adjust the position of other page objects. It allows print stream utilities to find the address block within the document and update/correct the mailing address without impacting other objects within the page. And, it allows the RIP to optimize how it renders the objects, such as pre-building the logo object once, caching it, and calling it whenever it is referenced.

The first generations of transaction documents, starting in 1958 with the introduction of the IBM 1403 impact printer and continuing to the introduction of AFPDS in 1984, were created in much the same way as the documents they replaced; they were typed using steel slugs. In countries using European alphabets the print file started on the upper left-hand corner of the paper and added characters (and spaces) until it reached the end of the text line, and then placed a carriage return command to move the print carriage to the next line. Sophisticated programs may have had skip to next page commands. The printers of that era used steel characters held together on a print-train. The fonts were hard objects that were hammered into an inked ribbon in front of the paper. If you wanted to change the font you could only do so before you started the print job (by removing the print-train and installing another). Early print files did not delineate the beginning/end of documents or even pages. To move the address block it might be necessary to shift multiple print lines, potentially misaligning all of the trailing information because it had to be exactly placed onto a pre-printed form. Although this may seem like ancient history, there are hundreds of 1403-style print files printed every month in large enterprises.

In a recent survey of its clients, Acadami found that approximately 90% of their clients' mainframe print jobs (by number, not volume) were described as either line-mode, 1403, or AFP Compatibility (line mode with pagedefs/formdefs).

Building with Objects

With the introduction of electrophotographic laser printers that could address any point on the page, the technical constraints of impact character printing were eliminated. The RIP could place information anywhere on the page as it rendered the raster pattern.

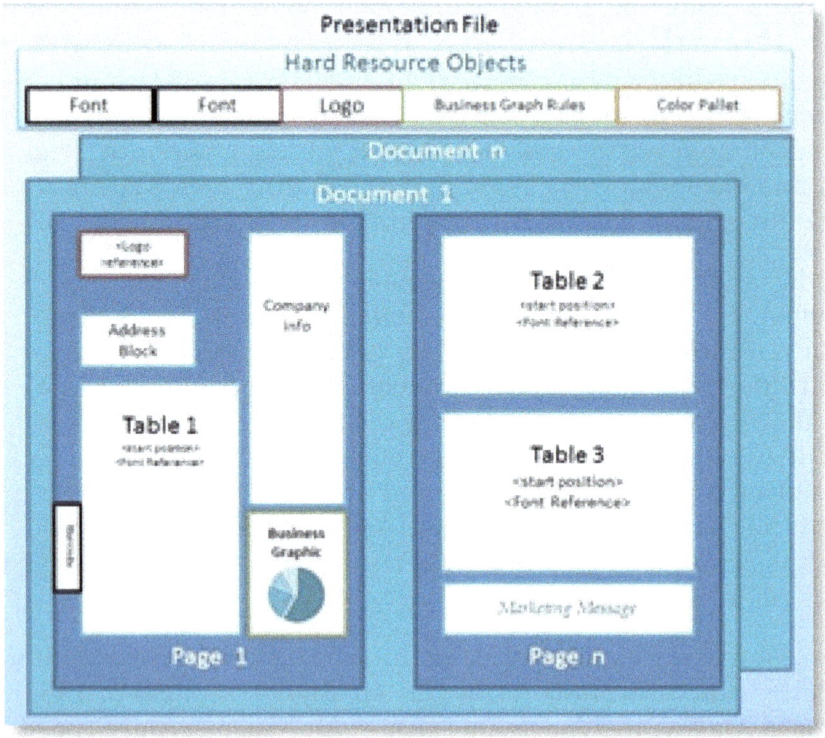

Figure 26 - Example of nested objects within a presentation file. Created by acadami.

By using multiple objects to build pages and documents, file sizes can be structured to minimize redundant instances of resources, saving significant space. Documents and pages are identified, allowing a print controller or a document viewing application (such as Acrobat) to interpret a specific document or page only, and its referenced resources.

By identifying hard or multiple-use resource objects at the start of the presentation file, the RIP only has to render repeated resources once, and then cache them for use when referenced. This speeds up the rendering process while allowing the RIP to better calibrate (e.g. ICC color profile[1]).

Text Block Objects are anchored by referencing a starting X-Y position. The position can be absolute (positioned from the upper left hand corner of the page) or relative (positioned in relation to another object). To adjust the address block object on page 1 in Figure 3, change its start position from the start of the page (an absolute positioning). Or, widen the space between the bottom of Table 1 (in Figure 3) and the start position of Table 2 by changing the relative position.

In the case of the Barcode object on page 1 of Figure 3, the object can reserve space around it. This keeps other objects from interfering with its quiet zone, an area surrounding the barcode object that is deliberately left blank to ensure a proper read.

1 This means that the RIP doesn't have to do a complex ICC conversion each time it is given a colour attribute.

Components within a Compound Object

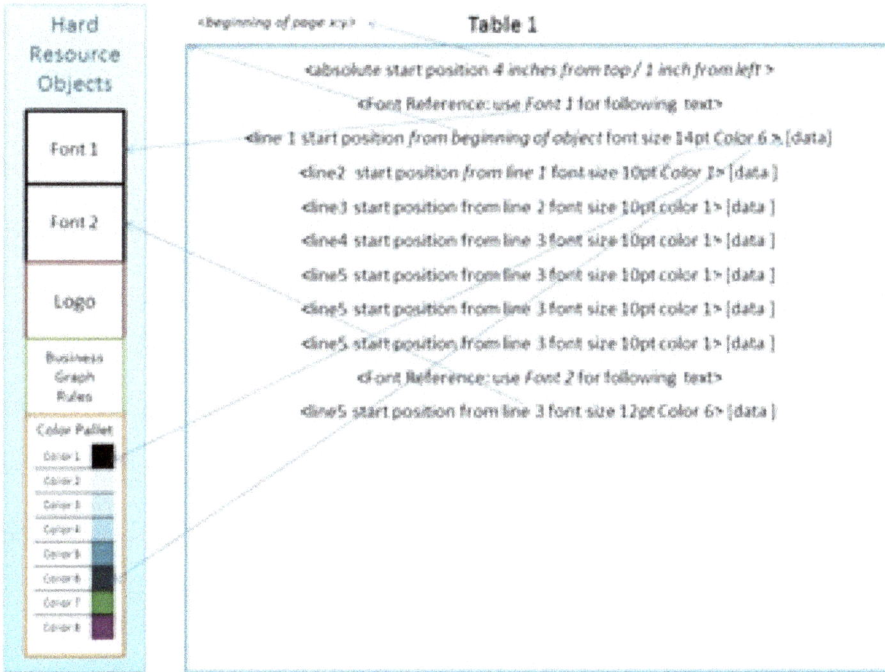

Figure 27 - Example of a Compound Object - Table 1 Structured text block. Created by acadami.

Compound objects should be considered unique entities. It should contain either all of the resource objects and data to be placed, or it should reference hard resource objects. A compound object should not reference resource objects found in other compound objects within the presentation file; e.g. Table 1 in Figure 4 should not reference a font object that was carried within the address block object. These objects may be replaced over time with unintended consequences.

A compound object will position the beginning position either absolute to the beginning of page, or relative to another object (usually the previous object's end position). It will also specify its orientation relative to the orientation of the page. For example, the compound object carrying the Barcode in Figure 3, is set at a 90° degree orientation to the page. All other compound objects are set at 0°.

A compound object positions information (Data) within the object. This can be absolute, as in Line 1 in Figure 4, or it can be relative, as in Line 2.

A compound object can reference a Font or Color Palette, either to be used within the entire object or by specific text placement. It can also carry these resources for its exclusive use.

Graphic, Barcode or Image objects can either be referenced or carried.

Types of Compound Objects

Adobe Portable Document Format (PDF)

Although we think of PDF as a Presentation Stream or Format, a PDF object can be carried within a page or document of a larger PDF file or an AFPDS file.

One of the strengths of PDF the ability to containerize resource objects and information, and to isolate hard resource objects at the beginning of the presentation environment (file). PDF actually carries most of its hard resources between page 1 and 2 of the file. It only carries the resources necessary to build page 1 at the beginning. This allows Acrobat to quickly render page 1 for the recipient, giving a respite to load the rest of the resources before the recipient views page 2.

Color Objects

Understanding Color

Color is the most difficult object-type to replicate, adding significant time and resources to both implementation testing and ongoing production.

Psychology of Color

Color emotionally affects us. Some colors, like red, make us more attentive. Others, like blue, make us more imaginative. Warm colors can make us energetic or happy, while cool colors can make us relaxed. The lack of color can cause depression, like Seasonal Affective Disorder, that is treated with the use of bright light.

Figure 28 - Microsoft Bing link blue drives more sales than Google link blue.

Microsoft proved that you can enhance or inhibit curiosity through the amount of red within the blue used on web links. They drove over $80 million in increased click-through revenue on their Bing search engine by shifting from reddish blue to greenish blue.

Pharmaceutical companies can demonstrate statistical differences based on the color of the pill alone. Although white is most commonly used (because we consider it pure or clean), warm colors are sometimes used to inspire hope. Light blue is sometimes used to signify calmness.

Corporate Identity

Organizations use one or two specific colors as part of their corporate identity, usually defined in a Corporate Style Guide . Much thought is put into the specific color combinations. What psychological message does it provide? Consider Starbucks green, intended to encourage customers to 'take a calming break'. For some logos the color has historical connections, such as Coca Cola red. Some logos do both, such as the IBM eight-bar blue on white logo, evolved from the Trusted IBM Salesman stereotype: blue suit, white shirt, and striped tie.

Corporate identity colors are used for more than the logo. They are also used to inform a variety of document elements, including:

- headings,
- call-out text,
- lines or boxes in tables,
- business graphs,
- line-drawings, and
- photographs, either as the:
    - shade color in a monochrome picture,
    - dominant color in a full color picture.

In situations where you see only one color used on a white background, the secondary color listed in the corporate style guide is often white. The guide may also allow the use of shades and tints of the corporate identity color; the color hue doesn't change but the luminance does.

Using Color Objects

Moving from black on white to full color printing processes creates the need to be aware of the different points in the process where color objects may be built, stored, re-profiled and otherwise touched. Here are some touch points:

- **Document object management system:** Manages color palettes and profiles.
- **Composition system:** May need to define color palettes and profiles for print or display devices.
- **Print stream utility:** May need to convert color definitions for finished product.
- **Print management system:** Must understand color attributes before assigning color print device.
- **Archive system:** Must be able to faithfully render colors as printed or displayed.

- **Electronic presentation system:** Must convert color profile for web presentation.
- **Printer:** Must convert RGB color profile to CMYK before rasterizing page.
- **Inserter:** May need to read color marks as part of manufacturing process, and could print colors onto envelopes.

The only steps that are color-blind are generally the data extract (ETL) routine and the postal delivery.

How Color Works

To appreciate the power of color, understand the science behind it.

Color Wave Lengths

Color is electromagnetic radiation somewhere between x-rays and radio waves (400 - 700 nm). It is also known as the visible spectrum. It's the tiny portion of radiation that we can actually capture with our eyes. Between 400 nanometres (violet) and 700 nanometres (red) are the color wavelengths that spread out into the spectrum of a rainbow.

Seeing Color

There is a matrix of photoreceptor cells called rods and cones in the back of our eyes. There are about 120 million rods and 6 to 7 million cones in each human eye.

The rods are one thousand times more sensitive to the intensity of light, but insensitive to color. They differentiate images as black, white and different shades of grey. They pick up luminance within color, such as black text on a white background or the shade / tint added to a color hue. They also become increasingly important in lower lighting.

Each of the cones contains one of three pigments sensitive to RED, GREEN or BLUE. Each pigment absorbs a particular wavelength of color. The short wavelength cones absorb blue light, the middle wavelength cones green light, and the long wavelength cones red light.

A practical application of this is that when designing transaction documents for low-light environments, make sure that the text is absolute black and the paper is absolute white.

Colors of Transmission – Red, Green, and Blue

Red, Green, and Blue are the three pigments in the cones of the eye and also the three primary colors we use to transmit color. When all three colors are transmitted onto a white surface, we should see white where they intersect. These colors are usually referred to as channels, e.g. the red channel.

Where only two of the three color channels intersect, there are the three color channels of absorption, Blue + Green = Cyan, Green + Red = Yellow, and Red + Blue = Magenta.

So, why are bananas yellow?

Bananas look yellow because their surface absorbs light-waves between 400 and 500 nanometers, which is the blue spectrum. The blue is removed by a yellow filtering process from the yellow pigments in the banana skin. As they go brown, the surface begins to absorb red and green.

RGB Values

Specific hues are achieved by adjusting the intensity of the color transmission. This can be described in two ways:

- In most publishing applications a hue is described by assigning a value from 0 – 255 (a byte of information) for each of the color channels, e.g. Cyan is R 000, G 255, B 255.
- In a web application a hue is described using the hex representation for each color byte, e.g. Cyan is: 00FFFF. This is called a hex triplet.

The Fourth Color: Alpha (white)[2]

The Alpha Channel is the 8-bit layer in a graphic file format that describes translucency. The additional eight bits per pixel serve as a mask and represent 256 translucency levels from entirely clear (0) to opaque (255), with levels in between representing the degree of haziness. RGBA uses a byte to describe the four color channels, e.g. R 000, G 255, B 255, A 127. Graphics in TIFF, PNG and TARGA formats support the 8-bit alpha channel; while GIF formats support a 1-bit channel to indicate transparency/no transparency. JPEG formats do not support transparency.

Colors of Absorption

Although we use light to transmit color, we use pigments to filter out light and reflect the remaining color. On printed documents, ink and toner reflect color because they contain pigments that absorb transmitted colors. This is known as absorption.

Although any color pigment will absorb light, we normally use permutations of three colors to create a wide range of hues: Cyan, Magenta, and Yellow. They are known as the primary colors of absorption.

The Fourth Color: blacK

Black is used to both shade hues (decreasing their luminance) and to provide an absolute black. The letter K is used to represent black, which is a legacy from the early days of printing when the black plate was used as the key for alignment.

Most of the readable information is printed in black. Using the black color channel to produce the black text will provide the best transition from black to white, making the document easier to read.

When setting the CMYK value for black text, ensure that it is “0, 0, 0, 100”.

Theoretically, you can mix cyan, magenta, and yellow to get black. In the real world, the mix does not fully absorb light and you end up with a brownish hue.

2 http://www.pcmag.com/encyclopedia/term/37669/alpha-channel

CMYK Color Values

Unlike RGB, we describe the amount of colour applied in each channel as coverage percentage. For example, Green would be described as C 100%, M 0%, Y 100%, K 0%, or “100, 0, 100, 0”.

Spot Color

Pantone

Pantone Inc. provides a process for standardized spot colors of absorption. Pantone colors can be specified to a commercial printer, an ink manufacturer, a paint company, or a textile manufacturer and what is provided should arrive with consistent color.

Pantone has introduced two separate color standards over its history.

Pantone Matching System

The Pantone Matching System (PMS) was introduced in the 1960s and uses 13 base pigments (plus 2 for white and black) to produce 1,114 spot colors.

It can be difficult to match PMS colors identified in corporate style guides. They are approximated with the CMYK printing process as the 15 different pigments do not directly map to CMYK. In addition, the exact formulation is Pantone proprietary information and can change with each type of target substrate.

Spot colors are optimized for solid areas of colour. They’re brighter and crisper than combined colors; it’s difficult to duplicate them with CMYK printing.

Most PMS colors cannot be matched in RGB, forcing most style guides to list an RGB approximation for web presentation and electronic documents. We can use the RGB value when profiling CMYK printing.

Pantone Plus

PMS was updated in 2010 with the introduction of Pantone Plus. It provides an additional 560 colors that are based on the same 13 + 2 pigment formulation.

Pantone Goe System

The Pantone Goe System, introduced in 2007, uses only 10 pigments plus clear coating for reflection. The new system was developed to be more compatible with color management software. Pantone has not matched the older PMS to the new Goe colors, stating that they do not always match. They advise organizations to continue to use the original PMS number for existing color specifications (such as a corporate logo) and use the Goe system numbers when picking a new color.

ICC Color Profiles[3] - Bridging between RGB and CMYK

We create color objects in an RGB environment but need to produce them in CMYK.

The RGB devices that we use are constantly improving in terms of the color depth, known as gamut, which they can display. Each display technology has significant

3 At the time of printing, the current version is ICC.1:2004-10 (Version 4.2.0.0)

differences in color channel transmission. They also degrade over time because of burn-in and other factors.

CMYK process color printing also has idiosyncrasies. Inkjet, electrophotography, and lithographic processes have different generic gamuts. Each supplier has their own tweaks to the process. In addition, the paper used will change the results based on color, opacity and gloss.

Although manufacturers provide ICC Color Profiles for their devices, many organizations use spectrometers to adjust the profile on a specific monitor or for a printer / paper stock combination.

The International Color Consortium (ICC) has developed a standard[4] for mapping a (usually RGB) device's color space using the CIE XYZ color gamut[5] as the connecting color space, and then using it to create a target profile (usually CMYK).

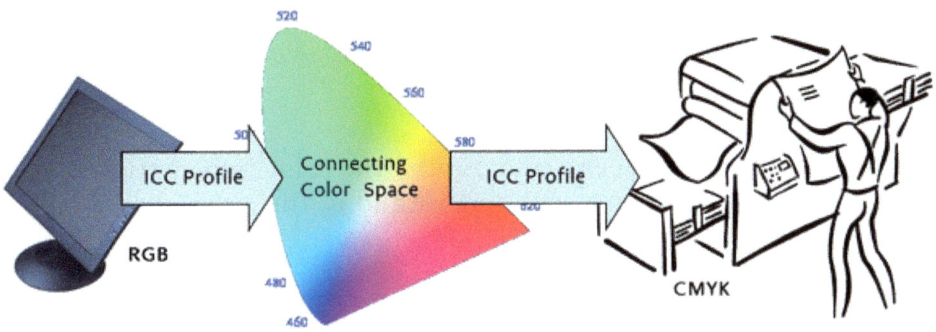

Figure 29 - ICC Profile Space.

Compensating for Color Blindness

Approximately 7% of the Caucasian males are color blind (and only 1 in 10,000 females). Color blindness is a genetic abnormality in the cones of the eye.

When planning to use color to highlight or direct the reader, add additional clues:

- avoid mixing pale green and red,
- have at least 60% contrast between text and background,
- graphics:
  - use Different line weights and line types (solid, dashes, dots),
  - use Patterns in color areas on bar or pie-charts, and
  - attach labels with leader lines.
- text:
  - italics,
  - bold, and
  - different point size.

4 Draft International Standard ISO 15076-1:2005.
5 Introduced by the International Commission on Illumination (CIE) in 1931.

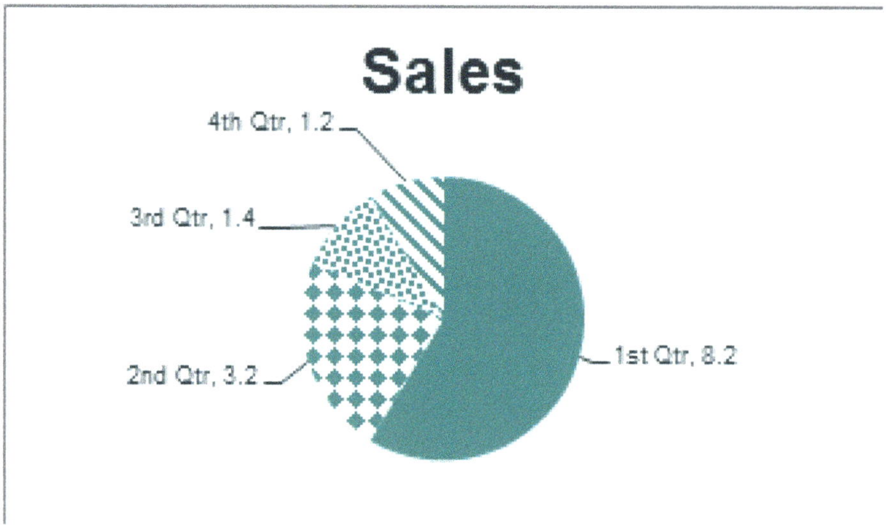

Figure 30 - Colored chart with additional color-blind clues.

The best way to see if a color-blind person can understand the color information is to make a black on white photocopy. If you can still understand it, the clues should be understood by someone who is color blind.

AFP Color Management Object Content Architecture (CMOCA)

"A resource architecture used to carry the color management information required to render presentation data... It defines objects that provide color management in presentation environments. These objects are called Color Management Resources (CMRs).[6]"

The CMR is imbedded within an object container, describing its color characteristics. These containers include the print file, individual documents, one or more pages, and specific objects such as images, graphics, fonts and barcodes.

CMOCA supports TIFF and JPEG image objects.

6 Color Management Object Content Architecture Reference AFPC-0006-01. http://www.afpcinc.org/site/assets/files/1068/cmoca_reference-01.pdf

Document Composition

Contributed by Carol Fiore, Roberta McKee-Jackson edp

What is Document Composition?

Composition involves processing all components involved in creating a document – raw data, document design criteria, business rules, static messaging, dynamic messaging, barcodes, postal sortation, rendering requirements – into formatted documents for delivery in print or via the wide variety of digital channels and media.

The composition step is at the core of the document production workflow and serves several purposes:

- It interprets formatting commands interleaved with text and graphics to create composed pages and documents.
- During the formatting process the composition engine may also be capable of interpreting business rules to build dynamic graphics, assign marketing messages, replace variable data fields with live data, insert customer support messages, and add compliance messages.
- It emits composed pages into one or more presentation streams suitable for print, archive and/or electronic delivery.
- It provides document integrity and workflow control data for managing downstream processing.
    - The downstream process management can be accomplished by various marks on the documents (barcodes etc.) and/or by control files that describe the presentation file generated in terms of documents, pages, inserts for each document, mail-packet weight, addressee, and other metadata.
- It can produce MIS reporting data on document content for CRM and effectiveness measurement as well as to meet audit requirements.

A well-constructed composition application, in conjunction with a quality document development project and appropriate document objects, yields well-formatted documents that process efficiently during printing, insert reliably, and ensure high-integrity end-to-end document manufacturing. Whether the documents are output to print streams or to display devices, quality composition facilitates faster, more efficient processing.

Composition History and Trends

Computer-based document composition has evolved along with computer technology, programming, and processing capabilities.

The earliest composition applications, developed at a time when there were mostly impact printers and they could only handle line-based presentation, used a variety of computer programming languages to insert formatting commands. Print files were limited to the typeface of the print chain on the impact printer. Changing fonts involved manually changing the print-train in the printer. The fonts had fixed positioning horizontal and vertical characteristics.

The combination of limited function impact printers and general purpose programming languages resulted in limited capabilities for composition. A wide variety of these legacy applications remain today simply because there is still a need for the application to run. These legacy applications are costly and difficult to manage for all the usual consequences of age: lost documentation, the original programmers no longer available, and even lost source code.

The next generation of formatters evolved separately on mainframes and servers. A program called RUNOFF emerged from MIT and was later modified at Bell Labs; it is still commonly found as TROFF and newer variations on UNIX systems. More common in the enterprise world was the mainframe SCRIPT program introduced by IBM as a method of adding formatting Control Words, commonly known as dot commands. Files marked up with control words were processed by a compiler to produce a formatted document.

SCRIPT has many variations, including Waterloo SCRIPT and a variety of PC-based programs designed to allow mainframe document files to move to the desktop computing environment. The basic markup language included options for changing fonts, adding line spacing, identifying headings of multiple levels in a hierarchy, and adding index and Table of Contents entries.

Tag-based languages followed Control Word-based languages in the 1970s with the introduction of IBM's Generalized Markup Language (GML). GML used intent-based markup so that final formatting rules could be maintained separately from the physical formatting rules. Instead of using Control Words to skip a line and indent, a macro was created with the desired formatting commands and then mapped to a tag, indicated with a leading colon or other tag indicator. As formatting requirements change the formatting can be changed in the macro without changing the document markup. The IBM GML product was marketed as the Document Composition Facility (DCF) and became the basis for the international standard SGML (Standardized Generalized Markup Language), HTML (HyperText Markup Language) and the host of XML (eXtensible markup Language) variations in the market today.

Some early composition engines focused on reading and manipulating data and placing results onto predefined locations on a page – often into boxes or spaces on pre-printed business forms. Many pre-printed forms later became static electronic forms called by the printer through commands in the print stream. That evolved into data merging with electronic forms, ultimately with conditional processing to permit formatting variations.

In the early 1980s, as the adoption of xerographic technology-based all points-addressable printers (e.g. Xerox 9700 and IBM 3800) progressed, visual design came into play. Instead of the typewriter approach, printers were able to control each dot on the page, as printer controllers became more powerful and added memory.

The Xerox Approach

On the earliest Xerox systems electronic forms were created on the printer controller, which took valuable time away from the intended usage of high speed production print. Intran Systems developed proprietary workstations using WYSIWYG (What You See Is What You Get) technology to create electronic forms which were then compiled to run on the printers. Intran created the electronic forms (.FRM), fonts (.FNT), logos (.LGO), and images (.IMG) supported on Xerox printers. Later Elixir, Tyrego, and Lytrod developed similar PC-based software to create the same type of resources.

The Xerox Integrated Composition Systems (XICS) emerged from Xerox, built on the base CompuSet language, to serve the full page/full document composition requirements that emerged. Xerox subsidiary Document Sciences took the XICS/CompuSet products through a series of enhancements creating Monogram and Autograph to support the development of documents. Document Sciences, as an independent company, created their follow on product called Xpression, an XML-based tool and provided migration paths to the new platform. Many XICS/CompuSet installations are still in production even though the product is no longer supported by Document Sciences, now owned by EMC.

The IBM Approach

The IBM mainframe approach led to the development of a set of products designed to create electronic forms (OGL: Overlay Generation language) and to add formatting to line data files (PPFA: Page Printer Formatting Aid) within the mainframe environment. The development of the AFP architecture (originally Advanced Function Printing, later Advanced Function Presentation) to generate the output to drive IBM printers created an environment for creating and managing documents, electronic forms and graphic resources. Print Service Facility (PSF) consumed AFP files to create device-dependent print files targeted to specific IBM print devices.

Other Composition Vendors

A new generation of composition products followed that could exploit graphic capabilities. Many included interactive design tools that allowed for the inclusion of objects anywhere on the output page, and provided an accurate rendering of what the page would look like. Products like Custom Statement Formatter (CSF), Lynx (now Pitney Bowes Software DOC1), and EZ-Letter found niches in the enterprise transaction market. Many customers use them today.

Another set of products began to enter the market during the 1980s to provide support for merging variable data and pre-composed document templates. Some were specific to an industry, including DocuMerge (Image Sciences, now Oracle) in the insurance and policy issue industry. Products of this type required sophisticated workflows, command of the document composition process, and an understanding of the final output stream formats.

Later programs emerged that transformed final format document print files to different formats and provided document re-engineering capabilities. This crosses over to the document composition environment, but is covered in the article on transform products.

Color

Use of color grew as two technologies converged. The first was the availability of color printers capable of printing at sufficient speed and low enough cost to make them worth considering for enterprise printing. The second was the rise of the internet and web-based information presentation. The availability of color gave rise to concerns about overusing color in business documents. The addition of color required additional training on the part of document designers and the addition of tools to generate color graphics and manage device-specific color requirements.

Post-Processing Functions

As the development of mail packages became more sophisticated the composition software added capabilities to add barcodes and other marks to be used in post-print processes during finishing and mailing. Most post-processing functions are based on the interpretation of business rules by the composition engine. Rules-based processing today is used to manage the insertion of variable data, the selection of images for inclusion in the document, the selection of marketing messaging and the management of excess available white space in the document.

Document Types

Composition today can be as simple as interpreting basic formatting commands or as complex as interpreting formatting commands and business rules. The most sophisticated programs may be linked to Customer Communication Management systems. Basic business documents generally fall into these categories:

- **Text-based documents** such as letters, notices, policies, proposals, documentation, and instructions.
- **Transaction documents** that communicate business content including account status, changes, ordering and services details such as bills, account statements, invoices, explanations of benefits, and bills of lading.
- **Promotional documents** that typically have more graphical content to focus on sales and increasing business such as offers, coupons, and brochures.

Another way to look at composition focuses on the application(s) used to create the document. In this view the completed document could fall into any category:

- **Batch, high volume** documents (often transactional).
- **Interactive** where a user selects a document template and responds to prompts to determine contents of a document, such as claims correspondence.
- **On-Demand** where the document application runs on a server after a user specifies data that determines the content, such as a price quote for insurance.

Composition Processes

The composition step is one component of the workflow to create customer communications. Within composition, there is also a workflow, whether in a new application or a batch job that runs on a regularly-scheduled cycle.

Assembly of Data and Document Objects

For composition in support of formatting transaction documents there are several steps in the workflow. It begins with the assembly of data extracts and document objects. For batch jobs on regular schedules there is generally a trigger to automatically set the process in motion to extract, transform, and load (ETL) the required data and document objects. For documents such as correspondence, which may be personalized interactively (for example by a customer service rep), data and objects may be obtained in real time.

Data extract files, often held on the mainframe or other client server, are readied for use. These extract files may include customer account information, applicable transaction records, related taxes and fees files including current updates to any taxes being applied, marketing content triggers, and external denominators file such as unit trust market prices or other data which will affect the customer document being produced.

All required document objects should also be available from the document object libraries. This includes fonts, images, any dynamic messages to be incorporated in the communication, and any static text objects e.g., terms and conditions.

Business Rules

During the development of the application business rules are created to ensure the consistent configuration of the document whenever it is recreated. The rules are based on the type of document, its purpose, the method of delivery, and its intended usage. They are used to set up a template for the layout of the document. The template identifies where to place each text block on the page, including its absolute or relative positioning. It also identifies where graphics and images will be placed in the document and if they will they be repeated within the document. An example might be a corporate logo placed on each page in a multi-page document. A repeated graphic or image may be referred to as a hard object.

Other business rules are conditional; they provide guidance for what to do when certain conditions exist. Examples of conditional rules include language preferences, customer preference for electronic delivery or printed delivery, or both, regional conditions, special needs or accessibility requirements, and page threshold requirements in a particular batch process.

Business rules also provide instructions on how to break tables at the end of pages or how to carry table data to the following page if a break is required. There are also rules on how to force page breaks to avoid widows (a paragraph-ending line that appears at the beginning of the following page) and orphans (either the opening line of a paragraph or heading that appears at the bottom of the page by itself) in the page layout sequence. Business rules are also used to manage output options, including rules surrounding sortation for printing in mail delivery order, routing to digital mailboxes and other ePresentment options, as well as to archives.

Calculations

As the data and document objects are being parsed through the composition system there may be calculations to generate values to be used during composition and in the reporting process. The number of pages in the document is a common calculation and takes into account the rules for sortation and separation of the document based on the expected print device and format for printing. For example, printing two-up duplex may require insertion of blank pages at the end of Customer A's statement to avoid printing any of Customer B's data on Customer A's statement.

The composition system also calculates page numbers within each mail-piece as well as the total number of impressions in a print run. Impressions are the number of images printed as opposed to the number of physical pieces of paper printed.

There are other calculations that many composition systems are able to perform that should be avoided. For example, do not calculate any legally relevant information in the customer document. Examples include column totals in a transaction document; the date of issue or due date; interest charges; or the yield on a stock portfolio. These types of data calculations should always be provided in the data stream extracted from the client data file.

An audit file is normally generated with each data file produced for the composition system. In that audit file you would expect to see the total number of customer records generated and perhaps the total number of transactions for the entire data set, among other audit criteria. Normal calculations include verification of the total number of customer records and that the total value of the transactions after composition matches the number in the original audit file. This type of calculation ensures the integrity of the composition process but is not calculating any legally relevant data for the customers.

As a general rule, multiple passes are required to establish the optimum layout based upon the calculations, and to resolve such variables as "y" in "Page x of y" footers. Dynamic generation of documents creates information that affects formatting, and sometimes alters the placement of text or an object elsewhere in the document. Widow and orphan control is an example. This is often referred to as "two-pass" processing.

Layout

Document layout commands in the composition system give the instructions on all details of the document generation. The components of layout include:

- **Orientation:** Documents are usually created in either portrait (tall) or landscape (wide) mode.
- **Beginning of page:** First pixel in upper left hand corner of page which becomes the key reference point for all other objects on the page.
- **Logical page:** A virtual page. The system can place two or more virtual pages into a print image, e.g. two logical pages on one physical page yields one impression.
- **Text blocks:** Areas defined in the layout for placement of specific information, such as a corporate contact details.
- **Absolute positioning:** Placing an object or text at a specific point on the page. Absolute positioning is based on the beginning of page. This is often used in placing address blocks.
- **Relative positioning:** Placing an object or text on the page based on the position of another object, such as the beginning or end of a table.

Address Block

In a transaction document it is imperative that the address block fits properly in the envelope window (if window envelopes are in use) or that it can be read by scanners to match the address to the envelop. For window envelop environments if the address block cannot be read or is misaligned in the window, it cannot be delivered.

Specific criteria are established by every postal authority regarding placement and format of addresses. The criteria define clear areas required within the window as well as font sizes, acceptable font styles, and maximum height and width when addresses are not in a window. These requirements enable mail-pieces to be read by the high-speed sorting equipment at the post office.

The address block may also contain additional postal information such as postal barcodes applied through the composition system.

It is important to test the position of the address block early in the development process. Many project designs neglect to test the parameters of the address block early enough in the document design process which can result in expensive delays or damages if the application goes live without the required testing.

Emitter

As part of the core composition process just described, interim, often proprietary, format files may be produced by the composition system. This is a step prior to creation of the actual print file. The interim format is tagged to support multiple output formats; the tags are generally xml-like to provide information on structure and layout. Sometimes the interim format produces specifications for that job which can then be processed by other applications to prepare the job for the emission step.

The data stream may be run through multiple emitters creating print files, electronic delivery files, special needs files, etc. and these processes may run concurrently.

At some stage in the overall process, the documents may need splitting and sorting, for example by delivery channel, special needs, or archive-only, and then those destined for print by postcode, mail-piece weight, and overall number of documents. This may occur upstream of composition, during composition, or downstream using file manipulation software. If it happens during the composition step, then it is often part of the emission processing.

The resultant interim files are then run through the appropriate emitters to produce the desired output files. Depending on the format being produced, separate layout rules may be required for each output type:

- print files may emit as AFPDS, Adobe PostScript, Xerox Metacode, HP PCL, PDF or an XML variant,
- special needs or special handling files may be produced as Braille Ready Format (BRF), large letter, or accessible PDF, and
- electronic delivery may produce Adobe PDF, PDF/A, XML files, and in some cases image format files.

Production and Audit

As part of the composition process, production and audit control details are created to verify the integrity of the batch process. These include:

- **Document Number:** A unique identifier assigned to each mail-piece within the job.
- **Spill Code:** A human-readable number for use in the event the (cut sheet) documents are spilled on the floor: an incremental number placed on each page in the job.

- **Page Number:** Numbers each page plus total number of pages in the mail-piece.
- **Audit Sum:** The sum of the closing balance in each document (mail-piece); checked against closing balance sum from the calculation stage to ensure all documents in the batch have been created.
- Barcodes and files:
  - **Postal: Four State barcode: POSTNET** barcode that is being replaced by Intelligent Mail barcode, or PLANET (Postal Alpha Numeric Encoding Technique) code.
  - **OMR:** Earliest form of marking to indicate page count and end of document, and to control insert selection.
  - **2of5:** Page sequence numbers.
  - **3of9:** OMR fields plus document number.
  - **2D barcode:** Contains considerably more information in a very condensed space.
  - **Insert control file:** Created automatically by composition system to provide list of documents with attributes which is sent to the server managing the insertion process.

Insert Control File

During the emitter phase of the workflow in composition, an insert control file (ICF) can be constructed by the composition software. This insert control file is of particular importance in the downstream processing of the print file. The ICF provides the insertion process with the following information:

- Mail-piece identifier, e.g. CBC (Customer Bar Code) in the mail-piece field,
- Number of pages for each mail-piece,
- Weight of each mail-piece taking into account number and weight of any intended inserts, the weight of the document based on the number of pages, and the weight of the envelope,
- Thickness of mail-piece based on the paper characteristics, how many pages of the document are to be included, and the number of inserts,
- Inserts to be called based on the barcodes,
- Postal tray / pallet ID, and
- Address block information.

The insert control file is a valuable tool generated from the composition system and is the key to a robust insert control process, and to overall integrity of the production process.

Composition Software Products

There are many vendor solutions for composition with varying capabilities. Sources of information are available offering various market analysis studies on Document

Composition and Customer Communication Management (CCM) today, including those from Forrester, Gartner, and Madison Advisors.

One trend in businesses is to consolidate the use of composition tools, although migration without a need to change or enhance processing is not commonly considered to be cost effective. If new composition software is to be considered, key features that should be considered include:

- end-to-end application capabilities,
- design ease of use (WYSIWYG, collaboration),
- enablement of business users, possibly through interactive access,
- data handling flexibility,
- re-use support for resources and application building blocks,
- application flexibility (support of batch, interactive, and on demand requirements),
- integration, and
- repository capabilities.

Composition Best Practices

The following highlight key functional considerations for best practices in document composition.

Data

- Good data is critical to the success of customer communications. Corrections or changes to data (e.g. customer address) should be made at source if possible, for example in the customer database, transactions database, or marketing database, to ensure errors are not repeated endlessly.
- Automatic triggers to begin the extraction process should be built into the application design. Best practice dictates that normalization and sorting of data is completed in the ETL (Extract, Translate, Load) step before transmission to the composition system.

Design

Accuracy, usability, and flexibility are key requirements when designing document resources and applications. WYSIWYG design is a must. Reuse of elements across document applications is important for development productivity, reducing test requirements and maintenance costs.

The quality of the document design guide is critical to generation of successful, well-presented, and effective customer communications. Development of a good quality, up-to-date design guide is important to the composition phase; a desktop publishing tool file is not enough. Documentation should be clear and complete, with business requirements detailed up front. It should describe layout and business rules for all applications, typefaces, weights, styles and point sizes, page size, logos, and many other aspects of the document appearance. Other considerations are cut sheet, simplex or duplex, page orientation, N-UP, imposition capabilities, and booklet support.

When using outside consultants, designers, or agencies, be aware they may not be familiar with the variations which may occur either in the application, in production printing situations, or with multi-channel output requirements.

Testing design layouts requires diligent due process. See the Test topic for more information.

Trends and Considerations

Composition has provided an essential tool in the format and delivery of messages, not only to print but to the many optional channels for output. The composition utilities have expanded their function to deliver formats to Personal Computers through digital communication devices, the ubiquitous cell phones, tablets, and now wrist watches.

In this process, the utilities have gone beyond entirely outbound technology, to include interactive processing. Their future is just beginning.

Print Streams

Contributed by William McCalpin m-edp

What are Print Streams?

A print stream is a file that contains a series of bytes that are intended to print or be displayed in a print-like setting. The print stream describes what text and/or images to place on paper or other physical medium. It also describes how the text and/or images are to be placed on the medium. The result of the print stream is ultimately intended for the human reader.

By "print-like", we mean oriented to the page. That is, while print streams like AFP and PostScript are clearly oriented towards the placement of text and images on paper, a print stream like PDF is generally displayed on a computer screen – yet PDF is also a print stream because the layout of the content presumes a page-like setting: there is a page size, there is an orientation, and there are other descriptors to facilitate print.

Presentation formats such as HTML and XML are not page-centric; therefore, while they contain presentation commands, they are not "print streams" because they are not laid out on a "page". When you print an HTML page, note that the HTML does not have content to the bottom of the "page", but is as long as it is. If you print a long HTML page it will print over several physical pieces of paper, and you have no control over where the page break is. So, HTML is a presentation format, not a "print stream".

In the extremely simple case, a print stream might be just one or more records in a file that contain only printable text, e.g., "ABC", "DEF", "HIJ". In practice, however, print streams are almost always printable text and images interwoven with commands that direct the physical or virtual printer to format and present the text and images in a certain way.

Raster Image Processor (RIP)

All printers except one are raster devices; whatever text and images are sent to the printer, the printer must convert to a string of bits that can be placed on the paper or medium as dots that are black or a color. Whether the printer is a laser (electrophotographic) printer, inkjet printer, electro-erosion printer, or some other technology, the content is laid on the medium as a string of bits that are repeated row after row – with one exception.

The one exception is a pen plotter, a specialized print device that uses a pen to draw text and images on the medium. Normally used for scientific mapping or engineering plans, the pen plotter does not convert text and images to bits; rather it responds to vector commands that tell the printer to set the pen at location X,Y, lower the pen to the medium, then move the pen while drawing in a straight line or curve. Since pen plotters are never used as transaction printers, we will not discuss them further.

All transaction printers have a component (whether in hardware or software) that is called the RIP. The RIP (pronounced "rip", like "to tear something in two", as opposed to pronouncing each letter as "R"-"I"-"P") is required to convert the print stream to the raster stream that the print or display device will actually place on the medium.

The RIP has a critical influence on the performance of the printer, since the bit-level arithmetic involved is computationally intensive. Indeed, in the early days of electronic printing, there were few high volume PostScript printers because of the complexity of the PostScript language. High-speed PostScript printers had to wait until the development of both high-speed CPUs and the ability to run multiple high-speed CPUs in parallel in the RIP, so that multiple pages could be ripped simultaneously.

Even a print stream like AFP, which is optimized for high volume transaction printers, can be built in such a way that the RIP runs much slower than normal. That is, while a single page image is created from a single and unique bit map (a bit of an exaggeration, but close enough for our purposes), there are many ways to build a print stream to create that unique bit map. Take the case of the simple string "ABC". It could be represented in a print stream in a variety of ways. Here are just a few:

```
(x,y location)ABC
(x,y location)A(x,y location)B(x,y location)C
(x,y location)A(relative move)B(relative move)C
(x,y location)C(x,y location)B(x,y location)A
```

While each of these examples could result in the same final bit map for the printer RIP, the amount of processing to create the bit map from these strings can vary. The first example is the simplest and almost certainly the fastest to rip. The third example is likely to be faster than the second because the relative move updates only one aspect of the current address, whereas the second example updates both the x and y aspects of the current location. And, it is a peculiarity of Xerox Metacode printers that in certain circumstances the fourth example might not print at all.

Therefore, the choice of print stream architecture as well as the quality of the software that generates the print stream is of great importance to high volume document production sites, because a "bad" print stream can seriously slow down a printer or the transform that converts one print stream to another.

Contents of a Print Stream

A simple print stream contains text and images to be printed on a medium and the interwoven commands that direct the printer on how to format these text and images. However, for historical reasons, the environment surrounding the print stream complicates this scenario. In fact, there are a variety of components that may be found in a print stream:

- text,
- images,
- formatting information for the text or images,
- page or vertical control information for the printer,
- printer feature instructions (change tray feed, switch from simplex to duplex, etc.),
- references to external document objects (like fonts and images), and
- containers that hold in-stream document objects (like fonts and images).

In fact, all text characters are actually references to an external document object – a font. The pervasiveness of text and fonts tends to lead the user to forget that a reference to a text character is, in a way, the same as a reference to an external image or other object. This fundamental requirement to deal with internal and external resources is a critical part of transaction printing, and is a differentiator between it and other forms of data processing.

Text

The text consists of letters or symbols from a writing scheme. In the case of English, the letters are "A", "B", "C", and so on, but could just as easily be Greek or Arabic letters, or Chinese or Japanese symbols. It is the ability for text to refer to external document objects called fonts that makes it possible for a single printer to display every language known to humankind.

Images

The term image in this section is a general reference to non-text items, such as a photo or line drawing or logo or signature. For transaction printers there are two types of images: raster and vector.

Raster images are fundamentally bit maps of the image; that is, a two-dimensional array of bits that represent the image. Raster images are often, but not always, compressed. There are a number of compression schemes that can sharply reduce the amount of storage that an uncompressed bit map requires, such as one dimensional, CCITT Group 3/two dimensional (used in faxes), CCITT Group 4 (used in scanners and archives), and JPEG (used in photographs). The choice of compression scheme is dependent on whether the image is color or monochrome, largely text or a photograph, or lossless or lossy (this refers to whether or not unimportant bits are dropped during the compression process).

Raster images are easy to process in the RIP, but the sheer size of the bit map may cause performance issues.

Vector images are images that are described by a series of vector commands. Instead of providing a bit map of a circle on the page, the print stream may contain a vector command that states "at location x,y, draw a circle with diameter Z and with a rule width of Q". This command would typically be just a few bytes rather than the thousands of bytes required to represent the same image as a bit map.

Since all transaction printers are raster-based, the RIP must interpret the vector command to make the bit map of the circle; however, because of the much smaller amount of data involved, the RIP process and subsequent printing may take a smaller amount of time than using the raster image.

The ability to use vector commands is one of the reasons that PostScript was so successful for the graphic arts market – the page designer could use tools to create hundreds of complex images for the document master, yet because these vector commands were each small in size, the PostScript files were not unduly large. It was the number of vector commands in the PostScript file that caused the slow performance of early PostScript printers, a problem that has long been addressed.

NOTE: We are using the term *image* as a generic term. However, different vendors in the transaction document space have their own terms, which may conflict.

For example, in AFP, an image is always a raster image, but a graphic is always a vector image. For Metacode printers, there was no vector support, so graphic and image referred to raster images that were handled in two different ways. Windows Metafiles can contain both raster images and vector images, which makes them closer to an AFP Page Segment (a container for both types of images).

Formatting Information

The page is a two-dimensional framework on which to lay out text and images. Either an explicit or implied location for any content on the page will be provided in the print stream. In the original Xerox Metacode model, the X and Y addresses referred to physical attributes of the printer – the scan and dot directions. The scan address was a function of how the paper moved through the machine – a scan line was the line of dots that the laser made across the page image on the drum before the drum rotated to the next line of dots.

However, all other print streams that have an architecture (besides line data) have virtualized the page so that the X and Y addresses are independent of the physical properties of the print device. In the case of AFP, there is the inline address and the baseline address, with inline referring to the direction that the text moves when one character follows another. This virtualization makes it easy to print all modern alphabets, since the inline direction is to left to right for Latin-based alphabets (and many others), right to left for Hebrew and Arabic, and from the top down for some Chinese and Japanese documents. Once the virtual page has been described, then the RIP takes care of the text placement.

Other formatting instructions include defining the inline and baseline directions (to use the AFP terms), defining which font is to be used, defining which color is to be used, defining the location and size of an image, and defining the dot resolution of the virtual page.

Page or Vertical Control Information

Nearly all print streams contain commands to set page breaks and location on the page. However, the original print format – line data – does not have a command set. Therefore, elements external to the actual text data were used to control page breaks and text placement.

In the case of IBM mainframes, there are carriage controls. These are a single byte at the beginning of each record that tells the printer what to do in terms of pagination and vertical placement. A skip to channel 1 may tell the printer to go to the top of a new page while a print and skip 2 will tell the printer to lay down the current record as a line of text and then skip two lines (leaving a blank line) before laying down more text.

In the case of environments other than the IBM mainframe, the print streams sit in an environment in which formatting elements from the typewriter are used: form feed, line feed, carriage return, and others. These elements were required to print line data (i.e., text) on simple decentralized printers, but they are largely ignored by modern print streams and serve primarily as delimiters between one record and the next.

Printer Feature Instructions

Printers have more features than just placing text and images on the page. For example, cut sheet printers often have more than one tray to feed paper from, so there are commands that may be external to the print stream to indicate which tray is the source tray. In the case of Xerox Metacode, the tray feed selection was done in the JDE (Job Descriptor Element) or by DJDE (Dynamic JDE) totally external to the Metacode environment. In the case of AFP, tray selection is done outside of a page or document object, but is part of the formdef, an external object to the print data. In HP PCL, the tray pull command is one of a group of escape sequence commands related to control of the paper (Esc&l#H).

Printer features and how they are implemented are highly dependent on the printer. In the first release of PostScript, because PostScript printers at that point were being used only to create master documents in low volumes, there was no tray feed command because PostScript printers had only one tray. And PDF, whose primary destination is to be displayed in the Acrobat Reader as opposed to printed on a printer, has no tray pull command at all, although some workarounds have been implemented.

References to External Document Objects

The section on Document Objects describes the many kinds and varied uses of objects external to the print stream. Some objects, like fonts and forms, are printable, while other objects, like AFP formdefs and color tables, don't print directly themselves but modify how other content is printed.

These document objects are often highly dependent on either the composition system (if they are source objects) or the printing system (if they are used in the printing process). For this reason, the way in which the print stream refers to these objects will also vary.

In PDF, the use of external document objects is discouraged, and in PDF/A, external document objects are forbidden by the standard. This is done because a PDF document is intended to be as portable as possible, and having external dependencies would greatly complicate the ability to accurately represent the document on any platform.

Containers that Hold In-stream Document Objects

Most print streams are also able to carry document objects in-stream as part of a print file, but the printer does not simply print this object - it stores it somewhere so that it can be reused. Given that host computers typically have much more storage and library capabilities than a printer, it is usually more efficient to store document objects on the host computer creating the print stream or the one sending the print stream to the printer than on the printer.

If there is a driver in control of the printer, as is the case of AFP, then the driver on the host has total control over what objects are stored on the printer. It is easy for the driver to query the printer to see if an object is needed and if so, to download the object to the printer in-stream.

In the case of HP PCL printers attached to Windows there is no single host in control of the printer, but potentially dozens or hundreds of computers creating PCL for that one printer. In this case, no software that creates the PCL can know what objects will be on the printer, so each PCL print job clears all erasable memory on the printer at the start of the job and downloads all objects needed for the print job in-stream. Depending on the number of redundant objects (like a company logo that appears on many printed documents) this latter method is much less efficient because all print streams have to carry the company logo.

There are methods to download objects to a PCL printer, but without a single host/driver controlling access to the printer, it is difficult to manage them, and as a result, most users do not bother.

Encoding Schemes

A print stream is a series of bytes organized into records and then into files. A byte might be the binary string of "0100 0001" or "1100 0001". But how do we know what the string of bits stands for? We know this by mutual agreement on an encoding scheme.

An encoding scheme is an agreement as to what letters, numbers, and symbols are assigned to which 8-bit strings. For example, the string "0100 0001" is defined to stand for the letter "A" (capital) in the ASCII encoding scheme.

There are two major encoding schemes found in transaction printing: ASCII and EBCDIC.

EBCDIC, the Extended Binary Coded Decimal Interchange Code, is used mainly on IBM mainframe and midrange computers. While these computers are numerically in a small minority in the business world, their central importance to the core data processing of most businesses in the transaction print space gives EBCDIC a disproportionate influence on the industry. That is, almost every large insurance company, bank, credit card processor, and the like has in the back IBM mainframes handling the bulk of the data.

ASCII, the American Standard Code for Information Interchange, is the encoding scheme used on almost every other computing platform on the planet. All Windows PCs and UNIX servers in a company use ASCII while the mainframes (if present) use EBCDIC, a situation that often causes confusion.

ECBDIC and ASCII both define all the upper and lower case letters (A-Z and a-z) as well as the digits 0 through 9. In addition, they both define a number of punctuation characters, but not always the same ones. For example, the standard EBCDIC definition has a symbol for the not equal sign (≠) while the original ASCII definition does not.

However, in electronic printing, the actual character that prints is a function of the font, and most fonts have been extended to include a variety of useful characters. But, since these are extensions, these characters were not always placed into the same locations in the EBCDIC or ASCII encoding schemes, making translations from one scheme to another problematic when nonstandard characters are used.

EBCDIC Print Streams

EBCDIC Text: The original EBCDIC print stream was just text. This text might be accompanied by typewriter-style formatting controls such as line feed, carriage return, and form feed. This format is quite rare, however. Suffice it to say that if such data somehow were delivered to an EBCDIC-compatible printer, the print stream might accurately print.

1403 Line Data: So called 1403 line data is a file that contains print data in EBCDIC and in which every record of print data is preceded by a carriage control. This carriage control would be stripped away by the spooler and substituted with a Channel Command Word (CCW) that would direct the printer's vertical motion after printing the line of text. Each line of text was limited to 132 bytes. While simple, the 1403 format was (and is) enduring because of the longevity of the 1403 printers themselves, as well as the variety of other impact printers that would take the same print stream.

3800 Mod I: The 3800 Model I was an electronic printer from IBM that was intended to be a replacement for impact printers like the 1403. Because the printer was an electronic printer, the print engine had the ability to change fonts. The 3800 Mod I print stream had a carriage control in the first position in the print record, and a one-byte font identifier in the second byte, followed by the printable text. The ability to change fonts was limited; there was a limit of four fonts available for a print job and the number of characters in fonts was limited, as was all upper case or all lower case characters. The maximum number of printable bytes on a print record was increased to 204.

3211 Data with Xerox DJDEs: 3211 was the name for a particular type of impact print from IBM. While the 1403 had a print train (the chain that contained all the available characters that could be hammered on to the paper) that was made to print a line that was 132 characters wide, the 3211 was made for print lines that were 150 characters wide. Thus, the maximum 3211 print record was 151 bytes – 150 printable characters and the 1 byte carriage control.

Xerox took this printer definition in the IBM spooling systems and created a series of "LCDS" (line-conditioned data stream) printers that were laser marking engines from copiers coupled with front-end computers that emulated the 3211 printer while adding a great deal of functionality. While the electronic 3800 Mod I printer was able to control only the vertical movement and the limited number of fonts through the print stream, the computer on the Xerox LPS (Laser Printing System) enabled the printer to store hundreds of fonts, forms, images, and logos, as well as enabling the print stream to dynamically change the print formatting page by page or even record by record. The Xerox printer – also called a centralized printer, or a Metacode printer – appeared to be an IBM 3211 printer to the spooling system, but its capabilities far exceeded those of any impact printer. With the right mix of forms, fonts, logos, and dynamic print commands, the Xerox Metacode printer was able to create a wide variety of sophisticated printed documents and drove the US insurance industry in particular for many years.

AFPDS and IPDS: As sophisticated as the Xerox LCDS environment became, the print stream was still fundamentally a line data print stream that laid data down a record at a time. A total architectural revolution was needed to make a print stream that was All Points Addressable. And so AFP (Advance Function Printing, later Advanced Function Presentation) was created. AFP was designed specifically to address a number of performance issues in the data processing environment, such as high speed printing, efficient resource management, and error recovery.

Furthermore, two parallel definitions were created: AFPDS (Advanced Function Presentation Data Stream) and IPDS (Intelligent Printer Data Stream). AFPDS was a standard definition that could be printed on all AFP-compatible printers, even with varying features. IPDS was the print stream that could vary slightly from printer to printer, and enabled the print driver (originally PSF: Print Services Facility) to translate the device-independent AFPDS that users created to a device-dependent targeted IPDS print stream that sent print data and resources to each printer.

Note: AFP specifies the architecture, protocols, objects and command set. AFPDS is a print stream conforming to the AFP specification.

These print streams are both page-oriented and object-oriented, for superior processing, resource management, and error recovery.

ASCII Print Streams

ASCII Text and ASCII Text with Print Controls: ASCII text was the first and most basic print stream based on the ASCII encoding scheme. We have included "text with print controls" here because it is virtually impossible to create text without some sort of print control to identify line endings. This form of print stream is still found in simple applications (Think: Windows Notepad).

ASCII Text with Escape Sequences: Early in the development of impact printers, the informal convention was created that printer commands could be smuggled into the text print stream via an escape sequence. A very early example using the Shift In/Shift Out bytes (x'0F'/x'0E', respectively) could be found on the Epson MX-80 dot matrix printer. The Shift In byte caused the dot matrix printer to print characters in condensed mode while the Shift Out byte caused the printer to print characters in expanded mode.

The formal escape character is x'1B' in ASCII, and many printer RIPs were coded to not print a character on this character, but to look at the following one or more characters as arguments for a print command. Indeed, this process became the basis of the most successful of the escape sequence print streams: HP PCL.

Escape sequence print streams (partial list):

- Epson MX-80
- Xerox UDK (XES)
- QMS QUIC
- IBM PPDS
- HP PCL
- Xerox Metacode

Page-oriented Print Streams: Just as for the EBCDIC print streams, the numerous ASCII escape sequence print streams were limited in that they were founded on a line data-oriented structure. Even though several of the escape sequence print streams achieved all points addressability via commands in the language, they lacked the robust character of page-oriented print stream architectures that combined objects, state tables, and other features of programming languages. The only similarity these print streams have with escape sequence print streams is that, except for Interpress, they are ASCII character-based.

These page-oriented print streams became very complex and were originally targeted to the graphic arts market. Performance was much less of an issue than the ability to draw complex graphics and documents with a minimum number of instructions.

Now, thanks to architectural updates as well as large advances in the underlying hardware in printer RIPs, these languages (except for Interpress, which is obsolete in the market) are widely available in the transaction document space.

Page-oriented print streams include:

- Interpress
- PostScript
- PDF
- The PDF family , including PDF/A, PDF/VT, and others

Sequential Print Streams versus Object-oriented Print Streams

The majority of print streams are sequential in nature. That is, the print stream is designed to be processed by the RIP by starting with byte #1 in the print file and proceeding sequentially to the end of the file. Attempting to jump directly into the middle of the file to start printing may result in incorrect output because a printer command executed on page 2 may affect all pages following, and if page 2 were not executed by the RIP, then pages 3 and following may print incorrectly.

For this reason, simultaneous parallel processing of pages for escape sequence print streams is usually not practical. Fortunately, the relative simplicity of the command sets for escape sequence print streams did not require parallel RIP processing to achieve high volume performance.

However, to go beyond the limits of escape sequence processing as well as to make better resource handling and error handling an integral part of the architecture, the concept of objects was added to several print stream architectures. AFP and PDF added the concept of a page, a document, tables, and other components that not only allowed the RIP to detect errors in the print stream, but also encouraged the reuse of content in the form of images, fonts, and text fragments.

AFP has a detailed object-oriented architecture wherein any piece of content can be in a container (object) of some kind. One way in which AFP encouraged parallel RIP processing was to create an Active Environment Group for each page, even if identical with every other page. For the most part, each page can be extracted independently of the print file and ripped. However, even with AFP, because of its long past, there are commands that can be used which ruin the ability to do parallel processing.

PDF also has an object-oriented format. The benefit of this architecture could be seen in an early enhancement of PDF which enabled PDF files to display the first page while the file was still being downloaded. Prior versions of PDF had internal object tables at the end of the PDF file, which meant that the entire file had to be downloaded to begin processing. When the Windows PDF driver was updated to be multi-pass ("optimized" in Adobe's terminology) so that these tables could be placed at the beginning of the PDF file rather than at the end, it became possible to process and display the first page while the rest of the file was still downloading.

PostScript is an exception to everything written above. PostScript is actually a programming language with if and loop statements. Although PostScript was not intended to be a general purpose programming language, it is humorous to think of your company's accounting being done by your PostScript printer's RIP.

Common Print Streams Today

AFP

AFP and its related IPDS were created by IBM in the early 1980s as a robust, state and object-oriented print stream architecture that could efficiently drive many different models of high volume electronic printers. The acronym AFP initially stood for Advanced Function Printing (in 1980), but was changed to Advanced Function Presentation by 1990 to acknowledge that AFP could be rendered on a monitor.

AFP is publicly documented while IPDS was not at first because it was intended to be used only by IBM products (the PSF driver and the Common Control Unit on the printer). The open nature of AFP had two advantages in the marketplace: (1) It alleviated the concerns that companies had for a proprietary print stream like Xerox Metacode, and (2) other companies were able to build products that built or manipulated AFP print streams, which increased the size of the market. IPDS was later made open as other companies signed licenses to build parts of the AFP/IPDS print environment alongside IBM.

AFP is state-oriented as well as object-oriented. State-oriented means that certain AFP structures must or may be followed by other structures. For example, the IMM structured field (Invoke Medium Map) may appear in a document object between the EPG (End Page) of the previous page and the BPG (Begin Page) of the next page. Finding the IMM structured field in any other location in the AFP document would trigger an automatic error.

Object-oriented means that the fundamental AFP element of the structured field can be organized into objects such as page segments, pages, active environment groups, image blocks, resource groups, font character sets, and code pages. Structured fields can be saved or manipulated in groups rather than always individually, resulting in a high level of reusability.

In 2004, IBM created the AFP Color Consortium to help the company expand the AFP definition on color. By 2006, the entire AFP architecture was transferred to the Consortium, which was renamed and incorporated as an independent non-profit called the AFP Consortium. The AFP Consortium, consisting of 35 vendors who create, manipulate, or ingest AFPDS or IPDS, is now an independent standards body.

Xerox Metacode

In 1977, Xerox Corporation released the Xerox 9700, a high-speed, cut sheet, duplex laser printer that was suited for the production of documents for the office. Xerox combined their high quality copier marking engines with a computer front-end that enabled a great deal of control over the print environment. Indeed, its capabilities far outpaced those of the contemporary IBM 6670 and 3800 Mod I printers, which were slow (the 6670) or intended to be a replacement for system impact printers (the 3800 Mod I).

The first 9700s (and the entire family) handled line data and were either attached to IBM mainframes emulating an IBM 3211 impact printer or were standalone units that read the print stream from a magnetic tape drive. This latter feature enabled the 9700 and its kind to accept print streams from a large variety of computers, such as Digital, VAX, Honeywell, and so on. The line printer RIP was powerful in that the computer front end to the 9700 enabled the storage of hundreds of unique overlays, fonts, images, and logos. In addition, the 9700 had tables called JDLs (Job Descriptor Language) that defined the order of the fonts, the page size, the orientation, how the incoming print record would be parsed, number of copies, unique changes per copy, and an almost bewildering array of functions. Furthermore, DJDE (Dynamic Job Descript Entry) capability was added so that the print stream could, by means of an in-stream DJDE command, alter most of the features of the print environment on-the-fly.

As powerful as the LCDS (as it was usually called) environment was, it could not achieve all points addressability. Xerox developed a mechanism to access the underlying metacodes by which the printer's computer sent data directly to the marking engine. These metacodes were simple – X and Y address, font choice, orientation, and relative movement – yet when created by the right software, they were capable of quickly printing high quality business documents. While the line printer environment imposed enough overhead on the on-board computer to slow the printer throughput to less than the rated 120 impressions per minute (for the 9700), the same print job in Metacode was normally able to run at rated speed.

The Xerox Metacode environment succeeded in business because a number of companies both allied to and independent of Xerox created software capable of producing high quality Metacode print streams. These environments – and the fact that the typical business document did not need the complexity of a document produced by the graphics arts – resulted in the widespread adoption of the Xerox 9700 line of printers as the de facto standard device in the US and Canadian insurance market – the largest market for these applications. Even though the Xerox Metacode architecture (although it isn't an "architecture" in the data processing sense) has been around since 1977 and is functionally obsolete, the quality of the applications is such that hundreds of companies still use Metacode printers and Metacode producing software. Until the advent of color there was never any requirement for business documents (as opposed to marketing documents) that the Metacode print stream could not handle as well as more modern AFP or PostScript print streams.

HP PCL

In the 1970s and 1980s a number of companies were developing their own table-top laser printing engines. Some of these engines were quite sophisticated and capable of full color, even in the early 1980s (e.g., QMS and its QUIC language). As there was no standard, each company developed its own print stream command set for its devices.

While the PostScript language could be printed on table top devices, the expense of the licensing fees and the slow processing due to the overhead of ripping the language caused it to fail to gain widespread acceptance for the printing of business documents, whose formatting needs were relatively simple.

PCL, on the other hand, was simpler to code. Furthermore, as an escape sequence language, its demands on the printer's RIP were limited as most of the formatting had already been done on the host device – the printer had to simply read the print stream and execute the simple commands such as X, Y location, font selection, lay down bit map, etc. Eventually, the PCL command set became quite large, making it possible not only to be all points addressable but to also describe any sort of business document in black and white or color.

In addition, the success of the Windows operating system coupled with the free Windows PCL print driver enabled every application that was Windows-compatible to be able to print documents on a PCL printer.

PCL is the most widely used escape sequence print stream, and is supported by hundreds of vendors. However, it has not enjoyed the same success as AFP in the high volume transaction document world because it lacks the architectural robustness of AFP, which was designed to be efficient and with good error recovery in high volume environments.

PostScript®

PostScript[1] is a programming language as opposed to a print stream; however, since it is not a general purpose programming language but one developed to support the particular application of printing, it is included with print streams.

Because PostScript is a programming language, it is capable of describing extremely complex graphic images in a relatively small amount of code. This was a positive for the graphics arts market, who wanted a powerful tool to build complex images, but it was a negative for the business world that needed speed more than complexity. Some early PostScript printers intended for the transaction document world would actually take up to twenty minutes to parse and render the PostScript file before it would print a single page.

PostScript was developed by John Warnock and Chuck Geschke, who had both worked at Xerox PARC (Palo Alto Research Center) on the Interpress language, but who both left when Xerox failed to develop Interpress as a successful commercial product. At first glance there is a high level of correspondence between the command sets of Interpress and PostScript, but PostScript was developed for the rapidly growing world of personal computing, so the initial implementation used printable text characters only, and so could be transmitted over any type of data line.

Hardware and software finally caught up with the CPU-intensive requirements of PostScript, making high-volume transaction document printers feasible. When high-volume inkjet printers with color became feasible, PostScript was already in place to support it, although there were still some performance constraints.

PostScript is still a proprietary print stream belonging to Adobe, and is the de facto standard for the graphic arts industry.

1 PostScript® is a Registered Trademark of Adobe Systems Incorporated

PDF

PDF (Portable Document Format) is a print stream that has been architected to display anywhere that its "reader" can run. PDF is a "print stream" because it is founded on the idea that there is a "page" of a certain size and orientation, and that text and images will be placed on that medium. While it is true that the normal medium for PDF is an electronic display, PDF differs from HTML and XML because its emphasis on the page allows it to guarantee that a PDF document will display AND print as accurately as possible no matter what platform is used. Compare this to HTML which cannot guarantee page parity or other print attributes.

PDF is able to perform its feat of document portability because the RIP is not in a printer but in a software product called the Adobe Acrobat Reader, which has been developed by Adobe for most platforms.

PDF is based on PostScript; however, it is not a programming language because it has been simplified to remove programming features, such as conditional processing and GOTOs.

PDF 1.7 was made into an ISO standard, ISO 32000-1:2008. PDF proved to be such a useful platform that a number of specialized versions of PDF have been developed. The versions of interest to the transactional document space are:

- **PDF/A - ISO 32000-1:** Designed for highly reliable long term archival.
- **PDF/VT - ISO 16612-2:** Designed for "variable data and transactional printing", although it is not widely used in the transaction document space.
- **PDF/UA - ISO 14289-1:** Designed for universal access (i.e., for people with disabilities who need assistive technologies to comprehend documents in a variety of formats).

Note that while Adobe has handed PDF over to ISO standards bodies like AIIM, Adobe will still continue to enhance the Acrobat Reader, which may have the effect of making it appear that there are enhancements to the print stream definition.

Note that most PDF definitions do not actually include new PDF commands ("tags"), but they may require the use of PDF commands in a certain order, they may require that certain things be required that are normally optional, and they may require comment tags to add additional information for use in readers other than Acrobat Reader. For example, the PDF/A definition removes certain features of PDF, such as JavaScript, as well as requiring that all external fonts and objects be stored as part of the PDF file (the PDF definition does not require that the Base14 fonts be part of the actual PDF file).

Transformation

Contributed by Chris Halicki edp and Linda McDaniel edp

What Is Document Transformation?

Document Transformation is the process of converting a document file from one print or view format to another. Examples include transforming an AFP file to PDF or a Metacode file to PCL. Transforms provide a variety of options. They make it possible to take a file composed for a legacy print file format and make it available online in a standard online format, such as PDF, TIFF or HTML. Transform software may also be used to standardize print formats across an enterprise. Print streams such as AFP, Metacode or PDF can be converted to print on alternative enterprise printers or on distributed and departmental PCL or PostScript printers. This is a common requirement after a merger or corporate re-organization. Many transform products also support manipulation, standardization, and indexing of a print stream before it is loaded into an archive, when moving documents from an old archive system to a new one, or concatenation of files prior to printing.

In a production workflow, transformation normally follows document composition prior to printing or archiving, or it can be called as part of archive retrieval to deliver the document in a format that is more user-accessible.

What Happens in a Transform?

Transformation software parses an input file and its associated print resources (fonts, forms, images, page layout and control files) and generates the desired output format. The generated output may include inline resources or may contain pointers to external resources. External resources may be created in advance or may be built as part of the transformation. Most software allows for some changes to the document pages, such as adding watermarks or repagination, as part of the transformation.

While transforms generally convert page objects in one format to matching objects in a new format, due to the differences among print and view data streams transforms cannot always convert to equivalent objects in the output. Different types of page objects may behave differently during the transform process. (Refer to the Print Streams sections of this book for more details on each of the formats discussed below).

Text & Fonts

AFP, PCL 6, PDF and HTML all support TrueType or Open Type fonts. In some cases you will be able to use the same fonts in the output as in the input; but, very often that is not the case. Xerox DJDE-conditioned line data and Metacode use orientation-specific bitmapped fonts, and older-style AFP files may also use bitmapped fonts. PCL 4 or PCL 5 files may use printer resident fonts.

Transforms may do an on-the-fly conversion from an input font to an output font (if the font is available) or it may use a font mapping table. Transforms may also convert text to images.

In general, using font mapping when it is supported allows the transform program to run more efficiently because the font characters do not need to be converted. However, depending on the selection of fonts available in the output format the fonts may not be an exact match and spacing issues may occur in the output.

	Bitmapped / Raster	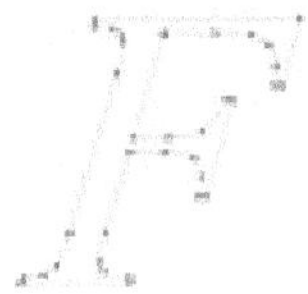	Outline / Scalable / Vector
	Xerox		Adobe Type 1
	Most AFP (C0...)		Windows TrueType
	Some PCL		Open Type
	Adobe Type 3		Most PCL
			Some AFP (CZ...)

If bitmapped fonts are automatically converted or rendered as images they may be scaled depending on the resolutions available in the input and output formats. Scaling may cause very thin lines and serifs to disappear and spacing may change slightly due to the differences in resolution if the new resolution is lower or not an even multiple of the original resolution.

Transform tools that offer font mapping may permit fonts to be mapped or rendered to an image selectively on a font-by-font basis. When choosing fonts for a font mapping table the best practice is to be aware of the character widths, the x-heights, the point size and style.

In the following example the text alignment variations are shown for 10 point Times New Roman mapped to 10 point Garamond or to 10 point Century Schoolbook:

Times New Roman

A line with some CAPITAL LETTERS and lowercase letters and some numbers 1234567890

Garamond

A line with some CAPITAL LETTERS and lowercase letters and some numbers 1234567890

Century Schoolbook

A line with some CAPITAL LETTERS and lowercase letters and some numbers 1234567890

The Garamond characters are smaller and narrower while the Century Schoolbook characters are wider than the characters in the Times New Roman font. Even though the fonts are quite similar and the point size is the same, the line lengths are different. The words and numbers shift to the right or the left. If each word was positioned exactly as in Times New Roman it could cause overlapping text or large spaces between words. If you are using transform software it is useful to acquire font creation tools or contract for services to ensure that you have matching fonts for your chosen output format.

Even fonts that have the same names or very similar names may not share identical x-heights, character widths or kerning characteristics. A Times New Roman fonts is not necessarily the same as the Times font or TmsRmn, even though they share the same design basis.

Encoding

There may be differences in the characters that are included in each of the input and output fonts, or differences in the encoding used in the input and the output. AFP and Metacode are often EBCDIC encoded, whereas PCL, PDF, PostScript and HTML are normally ASCII encoded. Most modern formats now support Unicode, but when transforming legacy formats, translation tables may be required to convert hex values to Unicode, or vice versa.

Some composition tools, especially print drivers, can use non-standard, dynamically encoded font subsets. In PDF files this is referred to as Identity-H encoding. Because the encoding of these fonts changes each time they are created, they can cause issues for transformations to fonts with standard encoding. Generally, identity-H encoded fonts will need to be transformed to another custom font or to an image.

Raster Images

All standard formats support raster images. There are usually few issues transforming black and white or single color images from one format to another. If the image is scaled and the resolution of the output format is lower than the resolution of the input format or the resolution is not an even multiple of the original resolution, very thin lines may disappear and slight spacing changes may be evident.

Images may be either inline in the data or stored in an external resource file that is called during page formatting. In a transformation, external images may be put inline in the output file, or maintained as separate resource objects. In some cases, such as transforms between AFP and Xerox Metacode, a separate transform may be executed on the image files to generate the new image files to be referenced in the transformed output. If transforming print files to HTML you may have the option to replace black and white images used for printing with color images optimized for viewing on the screen. If you are generating XML output, raster images may simply be ignored, and deleted from the output.

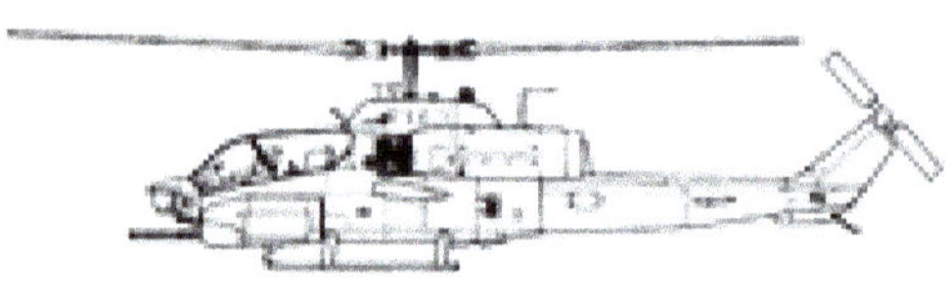

High Resolution

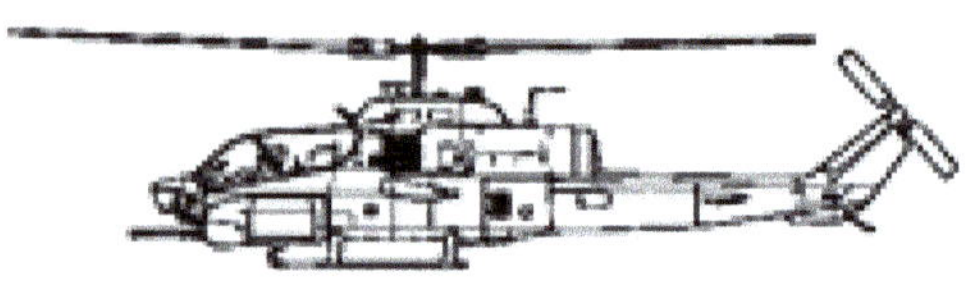

Low Resolution

Figure 1 - Effect of scaling caused by resolution differences.

Color Images

Color can offer more challenges, especially when converting a print format to a screen format, or a screen format to a print format. Color images for the screen are generally represented as a combination of red, green and blue light values (RGB) in percentages (0-1), decimal (0-255), or hex (00 – FF) values. Printed color images are usually represented as combinations of Cyan, Magenta, Yellow, and Black (CMYK). While there are standard formulas available for converting between RGB and CMYK color values, if the color temperature of the screen is not correctly calibrated or if the blend of the colorants in the ink or toner is slightly different, there may be a slight difference in the color that is presented in the transformed output. If possible, ICC color profiles (.icc) should be used to maintain color fidelity. If transforming from a color format to a black and white format for high-speed printing colored images may be dithered to create patterns of black and white dots to approximate the color shade. There are many standard algorithms available for dithering; if your transform software offers a choice, you may need to try several different options to find one that provides the best results with your images.

Vector Graphics

Line drawings, bar graphs or pie charts, or simple lines and boxes are often created using vector graphics. If your output format supports vector graphics (AFP, PCL, PDF, PostScript) the vectors are usually transformed to the equivalent vector commands in the output format. If the output format uses a different resolution than the input format, small openings may occur in the corners of boxes or in other graphics built from multiple individual lines. If your output format is TIFF or Metacode, the vector commands will have to be converted to raster (bitmapped) images. This can increase the file size of the output file. In some cases, multiple vector lines will be combined into a single image, but if the vector graphics are widely scattered on the page, it may be more efficient to create multiple smaller images in the output.

If converting to an XML schema confirm how the transformation handles graphics. Some transforms convert text but not the graphic elements, some require the graphic elements to be converted separately and linked back, and others simply ignore the graphic and image elements.

Shaded Areas

Shaded areas may be generated with either single color fills or fills created by repetition of a raster bitmap pattern. Shade fills may be transformed to equivalent instructions in the output format if equivalent objects are available. If converting to TIFF or Metacode, vector fill instructions will need to be transformed to raster images. If transforming to a format with a different resolution, bitmap shade patterns may end up looking like Op Art or basket weave patterns due to scaling.

When converting a print format to a screen format bitmapped fill patterns may be mapped to single color fills to look nicer on the screen and avoid the Op Art effects. Shading may be removed to reduce the output file size or to make it easier to read text in the shaded area.

50% B&W Raster Fill Checkerboard	50% B&W Raster Fill Scaled 80%	50% Gray Color Fill No change when scaled

When converting a screen format like PDF to a high-speed black-and-white print format you may need to replace color fills with hash pattern fills so that 2 different colors do not end up looking like the same shade of gray. PDF supports gradient fills, where the fill starts as one color on one edge and fades to a different color on the opposite edge. This type of object is only available in a few output formats so it may disappear or be replaced by a solid color fill in the new output.

Forms

Electronic forms contain content that can be used repeatedly in a print file even though it is sent to the printer only once. They go by different names in different formats – AFP Overlays, Metacode FRMs, PCL macros, or generally as Reusable Objects. In a transform they may be converted to an equivalent reusable resource (inline or external) or the instructions they contain may be "flattened" and repeated on each page where they are used. Some transform software requires matching form objects to be pre-created or pre-converted though separate processes.

Pre-printed forms will not automatically appear in transformed output since their artwork is not contained in the variable-data print file. If transforming to another print format, you can usually continue to print on pre-printed paper, but all spacing should be tested prior to production. If transforming to PDF or another viewing format you may have the option to add a digital copy of the pre-printed form, usually in JPEG or PDF format but other formats may be supported. Scanned images of the pre-printed pages are usually the least efficient way to add this information as the file sizes will be much larger with a scanned image than with a digital format that supports text and graphics.

Non-Printed Data

Input files may contain data that does not appear on the printed or viewed page, including indexes, AFP TLEs, NoOps, comment records, invisible text and other metadata. When there is non-printed data the best practices is to ensure that the purpose of the data is mirrored in the output file to maintain the integrity of the file. However, because non-print data do not affect the look of the output page many transforms discard it. It is common to see Comment or NoOp records transformed to equivalent non-operational data in the output format.

Non-printed data may also be used to generate index files as part of a transform process. Text in an invisible font (a font built with no bitmaps or character outlines) may appear on the output pages if the font is not mapped to an appropriate empty or white output font. Additional comments or NoOps may be added to the new output to indicate how the file was created or to document any changes that were made during any document manipulation steps.

Page Layout & Layered Objects

Page layout instructions such as margins and line spacing are normally preserved in a transform between print languages. Due to the nature of HTML page layout may not be preserved in transforms to HTML, especially if the HTML window size cannot be controlled.

Some formats allow page objects to overlap. Depending on the format the top layer may totally cover the underlying objects, but in other formats, the background is transparent, rather than white, and underlying objects may show through. Transforms may flatten all page objects into a single layer, or may group like objects together in the output – first text, then images, then forms, for example. This may cause text to be covered by an image, or text to be uncovered if the images are laid down first. Transform software may have options to maintain original order or optimize order by grouping objects.

Processing Instructions

Print files may contain printer and processing instructions, such as paper tray calls, duplex printing information, and binding instructions. If transforming to a view format like PDF, HTML or TIFF, these printer instructions may be discarded (screens are usually not 2-sided). For transforms to a print format, they may be automatically transformed to equivalent processing instructions, but they may also need to be set up in the transform configuration file, such as mapping "main" to "tray 1" and "aux" to "tray 2. Transform parameters may also be used to define what page size is associated with a page tray call, such as specifying that tray 1 contains Letter-size paper or tray 2 contains A4 paper. This information may be needed to generate appropriately-sized pages in output formats like PDF or TIFF.

Transforming for Online Viewing: Orientation and Resolution

If you are transforming the document for on-screen viewing instead of print you must consider orientation and resolution.

The majority of our documents are created in the portrait orientation. Most of our screens are in landscape orientation. If you shrink the page down to fit the entire page on the screen, it may be too small to read. If you display the same layout at a readable size, users will need to scroll to see some of the page contents. Alternatively, you may decide to redesign the documents, so that the online version does not exactly match the printed document, but still provides the information that your customer needs. In some industries, there may be compliance issues if the online version is dramatically different from the printed version.

A simple conversion of landscape pages may actually display them sideways on the screen (from top-to-bottom, or bottom-to-top) on a portrait page, and tumble-duplex print may display the backs of pages upside down. Transforms will normally need to be configured to rotate pages to make text readable on the screen. If some pages in the file are landscape and others are portrait, the transform software will normally need to analyze the direction of the text on each page, so that each page can be rotated appropriately for on-screen viewing.

As you can see in the table at the right, even our highest resolution screens barely come close to the resolution of our lowest resolution printers. Screen pixels are easily double or triple the size of printed dots. To represent on a screen exactly the same number of dots as on a printed page would require a screen 2 to 4 times the size of the printed page. If a PDF or TIFF file is scaled to the size of standard laptop screen, fine lines and fine details on images can disappear, or become proportionally thicker, due to round-off of fractional dots.

If input documents normally print on preprinted paper, matching images will need to be added to replicate the look of the paper on the screen. The transform software may be configured to bring in a JPEG or other image formats for company logos, text for forms, or add colored backgrounds. A full page PDF of the preprint design may also be used.

Dots Per Page

Screens		Paper	
CGA - 320 x 200	= 64,000	US Letter@ 240 dpi	= 5,385,600
VGA - 640 x 480	= 307,200	US Letter@ 300 dpi	= 8,415,000
XGA - 1024 x 768	= 786,432	US Letter@ 600 dpi	= 33,660,000
SXGA - 1280 x 1024	= 122,880	US Letter@1440 dpi	= 193,881,600
UXGA - 1600 x 1200	= 1,920,000	A4 @ 240 dpi	= 5,567,104
HD - 1920 x 1080	= 2,073,600	A4 @ 300 dpi	= 8,699,840
QSXGA - 2560 x 2048	= 5,242,880	A4 @ 600 dpi	= 34,799,360
		A4 @ 1440 dpi	= 20,093,952

Transforming for Enhanced Accessibility

When transforming to PDF, you have the option to add features to your documents to make them more useable. Text strings can be linked to URLs to allow users easy access to additional information provided on websites (This can also be done in HTML pages). Bookmarks can be added to a PDF document to provide quick access to selected sections of the documents.

But more importantly, documents can be transformed to PDF/UA, which allows them to read correctly by screen reader software, making them available to visually impaired customers as well as sighted customers. Transform software can be configured to automatically tag headings, tables and lists. Alternate text can be added for images and correct reading order can be assigned.

Transforms can also be used to generate Braille-ready format used by dynamic Braille displays or Braille printers for visually impaired customers who have been trained to read Braille. This type of transform may be triggered based on a customer's delivery preferences stored in a company's customer database.

File Size Considerations

After a transform the file size will not be the same as the original. It could be as small as 1/10th of the original size or as large as 15 to 20 times the size of the original input. There are two major reasons for the differences – compression and resource placement. For example, if you transform a single AFP (ACIF) file with all resources inline into a single PDF file (with all resources inline), the PDF file is likely to be about 50% of the size of the AFP file because the Flate compression used by PDF is a very tight compression algorithm. If you transform the same AFP file into multiple PDF files, as occurs when you convert a complete print run into individual statement documents for each customer, the total size of all the PDFs put together is likely to be many times the size of the original AFP file, because resources such as images, forms, and fonts are needed for each individual PDF file rather than stored once and used multiple times. File sizes can also increase if text or vector graphics are transformed to bitmapped images, such as occurs in transformations to TIFF.

Some transform software allows you to tune the output files size by setting parameters for image DPI, vector/raster formats, compression options, and resource placement (external or inline, on each page or as reusable objects). File sizes can vary greatly based on font selection, the number and size of embedded images and color fills, and the use of reusable or external objects. Always make sure you run a variety of tests with your own data to determine the appropriate amount of disk space needed for your transform output.

Planning a Printstream Transformation

There are three main steps to consider in setting up a print file transformation:

1. Determining What You Have.
2. Managing the Resources.
3. Adding Value to the Output.

Determining What You Have

The format of the print file will vary depending on the composition system you are using and the brand and model of printer the file was designed for. While IBM/InfoPrint/Ricoh and Canon/Océ printers normally accept AFP and PDF, Xerox printers may accept Metacode, PostScript, PDF, PCL or even AFP, depending on the model. You might even be using line data or a variation of line data such as Xerox LCDS or AFP conditioned line data. Your input may be encoded in EBCDIC if it is created on a mainframe or in ASCII if it is created on a workstation or network server. There are also variations within the major print formats. AFP can be well formed MODCA-P or output from ACIF, or it may be generated by programs that do not create well-formed AFP. The AFP may also have TLE indexes. Xerox Metacode can be designed for printers running online or offline, and may be mixed in the same file as LCDS. PCL could be version 3, 4, 5 or even 6. PostScript can be level 1, 2 or 3, and can possibly even contain inline programming. Even PDF has many variations. To determine what you have, first check the output options settings in your composition system.

If this is legacy data, that may not be an option. Your document designers or printer operators may be helpful in filling in the details. Alternatively, you can try opening the file in a text or hex editor. If everything in the file is readable in a text editor, your file is probably line data. If you see $$DJDE or $XEROX$, followed by a mix of readable and unreadable characters, you probably have Xerox Metacode. If the file starts with %%PS it is probably PostScript. PDF files normally start with %%PDF. If there are unreadable characters, switch to a hex editor, preferably one that will also show you both ASCII and EBCDIC as well. If the first character is a hex '5A', your input is probably some flavor of AFP. AFP records normally start with hex '5A', followed by a 2-byte length field, followed by a hex 'D3'. If the starts with hex '1B 45' it is probably some level of PCL.

Managing the Resources

The next step is finding the print resources and determining how they will be handled by the transform. Print resources are all of the other files or objects used by the printer to resolve the document, such as fonts, images, electronic forms or formatting files. Most print formats use a combination of inline and external resources. Transforms need to have access to these resources to get the best results.

AFP resources are normally stored in mainframe resource libraries, and sent to the printer with the print file. AFP resources include coded fonts (X0... or XZ...), code pages (T1...), character sets (C0... or CZ...), page segments (S1...), overlays (O1...), form definitions (F1...) and page definitions (P1...). The file names are normally limited to 8 characters. Recent AFP files may also use Open Type fonts.

Xerox Metacode and LCDS print files normally use resources that are stored on the printer. The file names are limited to 8 characters with a 3 character extension. Xerox resources include fonts (.FNT), images (.IMG), logos (.LGO), forms (.FRM), and JSL Formatting Files (.JSL). When the JSLs are compiled on the printer, other resources can be created like job descriptor libraries (.JDL), print descriptor entries (.PDE), and copy modification entries (.CME).

PCL often uses fonts that are installed on the printer at the factory and not available to the user. Different models of printers may have different standard fonts installed, making it difficult for a transform to accurately emulate the printer output. All other PCL resources are normally found inline, but occasionally external macros are downloaded to the printer prior to print time. PostScript files also use printer-resident fonts and may use internal or external macros.

If the resources are available, the document transformation engine can normally convert them to matching PDF objects on the fly. However, converting fonts on the fly can be inefficient. To solve the problem many transform programs use font and resource correlation tables.

In a Correlation Table the user can define what output fonts they want to use to match each input font. If a PDF input file uses Helvetica, Times-Roman and Courier (and their bold, italic, and bold-italic variations) the fonts can be mapped to the standard PDF Base-14 fonts. If the input file uses older fixed pitch fonts or more decorative fonts chosen, matching TrueType or Type 1 fonts should be found (purchased or created) so they can be embedded in the PDF file by the transform. That way the fonts will be available for proper viewing of the PDF. This will increase the size of the PDF file, usually by 20-30KB per font.

Similar processes apply to other transform pairs.

Adding Value to the Output

When converting a print file to PDF it is possible to add resources to make the document more useful and appealing to the end user. For long documents bookmarks can be used to create a clickable Table of Contents for the PDF. URL links can be added to create links to the company website.

For transforms to PDF and some other file formats it is usually possible to change font color and background color as well as to add or delete images.

If inserts are usually included in the envelope with the printed document most transforms support the ability to selectively add new pages to the file to permit replacement of those inserts. It is usually possible to add or delete text, such as disclaimers or watermarks. Most transforms support the ability to mask personal information like Social Security Numbers or an account numbers.

If the documents are going to be loaded into an archive or ECM system most transforms support the ability to output separate files for each account or customer, and to create an index with names, dates, account numbers and other information that can be used to load the print or PDF files into the archive and make their retrieval easier. If there are redundant pages in the input, like the "How to balance your checkbook" pages in bank statements, or blank or banner pages, they can usually be removed.

Consider all your options when planning a print file transformation. The more you know about your input data, and the resources it uses, the better off you will be. When configuring the transformation using a set of sample data that accurately represents all the variations in your input data will help prevent issues from arising when you move to production. Think about what how the final files will be used, and what might need to be added or deleted to make the file appropriate and valuable for the user.

Preparing for Transformation

When you have decided that a transformation is what you need to move a type of file or data from one format to another you will have to do some investigating into your own current processes and where the transformed data will be used. If the newly transformed data is going into a different workflow or the same workflow, there will be considerations regarding file sizes, resources and even if the process can interpret the incoming new format.

Investigate How the DATA Currently Prints

The first thing to do is trace the entire path the data takes from generation to final output on a printer or an archive. This means following the path the data takes from data entry, extraction from a data base, or composition system through any tools or utilities, no matter how simple. This can include file transfer, printer controllers, other transforms, or any in house tools for carrying the print towards its final goal. All of these can be adding to the original data.

There are many reasons to do this tracing.

- If your print data is part of a legacy system that may be generated by COBOL or similar programs on a mainframe it may not have been touched since it was designed. This increases the odds that there is some post-processing to update or enhance it.

- It may already have some form of transform or secondary program used to alter or manipulate it to work with more complex printers or finishing equipment. Learning what manipulation has already occurred to this data will help define what needs to happen with a transform you will be implementing.
- Learning about these tools will also help in ensuring that the format you create will work with existing tools. For example, you may have an indexing stage that requires TLE index data in an AFP (AFP Data Stream) file. It is certainly better knowing that before you discover the data is not indexable at the next step.
- Learning how the data prints and what it looks like prepares you for other questions. The data may be printed on pre-printed forms (roll or cut sheet). Do you need to mimic the preprinted forms in your final transform? Do you need to be concerned with tray or bin paper pulls in emulation of what you are doing currently. If going to an electronic format, you should consider that you might want to emulate the pre-printed forms with electronics overlays.
- It will be easier to answer vendor questions about your data as they should be asking these same questions.

This review will ensure that you know what the final output should look and how it should work in the existing system or adjust for a new system.

One other reason for following the path the data takes is the potential for an intermediate step to manipulate the data. It might be simply assumed and no longer thought of as a step. Especially in the case of legacy data, a simple program might be just a job step in your MVS JCL but it may be changing all the carriage controls to fit a target print. You might have line data on a Z0S or mainframe system that is in EBCDIC. If the data is moved to a server or workstation, it might be translated to ASCII. If that is what is being sent to the printer, that is what the transform will probably process. If the transform receives the EBCDIC data, changes would be needed to adjust to that format. If the data goes through a print controller, the format of carriage controls could be affected and changed to match a target printer. If the transform receives data, from before the control changes it, it would not transform in the same way.

Determine the Data Format

The print data you have could be one of many formats. Making an assumption that is it some form of Xerox format because it is intended for a Xerox printer might be a mistake. There are many printers that might have an interface that translates the data for that device. The Xerox printer could take in AFPDS data. To be sure, you will need to examine the data, preferably with a hex viewer. You should have a hex editor available to you when determining what sort of print data you have. This is because data can be EBCDIC, ASCII or both and may consist of readable text or data in an encoded format or both. The following section has some methods to identify them.

Formats

Print data can be in different formats. The most common are the following.

AFPDS formatted data or AFPDS: This contains x5A carriage controls in the first byte of each record. This data can have indexes called TLEs. The text data is often in EBCDIC but it can be ASCII if it is generated by a composition system. Often this data has been processed by an IBM Utility called ACIF (AFP Conversion and Indexing

Facility.) which also can ensure that all of the resources are in line with the print data. Compositions systems also often insert the resources inline.

You can see EBCDIC line data that is printed by Ricoh/InfoPrint printers through the use of a pair of print objects called a formdef and pagedef. If you see EBCDIC line data in an environment with AFPDS printers, the JCL that routes the data to the printer will show these two resources. Most transforms can process this line data. AFPDS can be printed on many brands of printers depending on the printer controller.

No matter what type of data, the print instructions are sent with the print data to the final device. They may be inserted by a composition, transforms or by the printer management software.

Xerox print data: Xerox toner printers have a hard drive that is used to retain resources such as fonts and forms. These printers use a source file called a JSL which is compiled into a series of objects that contain printer definitions such as font lists or job instructions. When transforming Xerox data of any format, the JSL used on the printer should be provided. Some transforms can read the compiled objects, but most can read the source files.

Xerox printers can take three formats of data natively.

Metacode. Move to the middle of the file in a hex editor and look for ASCII text. If you see ASCII text and instead of simple blanks between words you see hex values like x06, then the data is full composed Metacode. The print file might start with EBCDIC data that passes information to the printer but that same data is informing the printer that the rest of the file is ASCII text and formatting code. The beginning of the file could be ASCII also, in which case all the text is ASCII. If after reviewing the file you cannot find any readable text after the beginning, you should investigate to see if the file was transferred to a workstation as text and translated. If so, it is not usable as the encoding values are now useless. The sample below is showing Metacode ASCII text. The values between the words are command metacodes for spacing and fonts.

```
0D 09 4A 61  6E 75 61 72  79 13 31 31  2C 13 32 30 ..January.11,.20
30 38 13 06  3C 06 04 18  0A 00 01 0D  09 50 6F 6C 08..<........Pol
69 63 79 3A  13 32 34 4D  54 31 39 38  32 2D 34 13 icy:.24MT1982-4.
01 00 5F 00  00 01 06 8C  01 04 A4 09  00 02 0D 09 .._.............
54 6F 3A 13  06 B8 02 04  A5 09 00 01  0D 09 4D 6F To:...........Mo
65 13 54 6F  72 6D 6F 75  74 68 13 06  B8 02 04 6C e.Tormouth.....l
09 00 01 0D  09 39 13 44  65 4C 6F 72  65 61 6E 13 .....9.DeLorean.
44 72 13 06  B8 02 04 32  09 00 01 0D  09 44 65 74 Dr.....2.....Det
72 6F 69 74  2C 13 4D 49  13 34 38 37  35 35 13 01 roit,.MI.48755..
```

LCDS or Line Conditioned Data Stream. The data will most often be EBCDIC on a mainframe. The data will derive formatting information from both the job started on the printer (the JSL) and from DJDE commands in the print file that can format the data and invoke forms. This data is most often fixed length. If the data is ASCII, examine it in hex and see if there are any carriage return/line feeds, which are hex values x0D0A, after each line (x0A for UNIX). This will mean the data is translated as text during a file transfer. This might work with a transform, but since it is not the original format it is strongly recommended not to use it as any inline graphics or font index values that are based on half bytes will be destroyed. The example below shows an example of a LCDS file starting with printer instructions.

```
4E 40 C4 D1  C4 C5 40 C6  D6 D9 D4 C1  E3 7E D7 C4 + DJDE FORMAT=PI
D4 C1 D3 F1  6B C4 E4 D7  D3 C5 E7 7E  D5 D6 6B C6 MAL1,DUPLEX=NO,I
D6 D9 D4 E2  7E D7 D9 C7  F1 F9 F9 6B  5E 40 40 40 ORMS=PRG199,;
40 40 40 40  40 40 40 40  40 40 40 40  40 40 40 40
```

Line data. Plain line data is simply that. Usually EBCDIC, with fixed record lengths on the mainframe. All formatting is done by instructions loaded on the printer in a JSL source file. When a job is released from the print queue/spool, a job started on the printer will contain all the information needed to format. Below is an example of EBCIDC line data (in a hex viewer) or line data.

```
40 40 40 40  40 40 40 C5  86 86 85 83  A3 89 A5 85        Effective
40 D1 C1 D5  40 F1 F0 6B  40 F2 F0 F1  F0 40 40 F1  JAN 10, 2010  1
F2 7A F0 F1  40 81 4B 94  4B 40 A2 A3  81 95 84 81 2:01 a.m. standa
99 84 40 A3  89 94 85 40  40 40 40 40  40 40 40 40 rd time
40 40 40 40  40 40 40 40  40 40 40 40  40 40 40 40
40 40 40 40  40 40 40 40  40 40 40 40  40 40 40 40
```

PCL Print Data: PCL is ASCII based and often begins with ASCII text saying: @PJL ENTER LANGUAGE = PCL. All PCL commands start with an ASCII hex value of x1b, which is called an "escape character. When viewed in hex you will see this value often. PCL can also contain HPGL (Hewlett-Packard Graphics Language) commands which are derived from the HP plotter printers. This language can be used for fonts, but is most often used for graphics. Files often contain both formats. PCL generally contains the graphics and font data needed to print except for one important distinction. PCL printers of any type have built in scalable fonts. The system building the print data may be aware of those fonts and will not include them into the print file, simply refer to those fonts. This is the normal method used by Windows print drivers. Transformation software will need to address this situation.

Below is a sample of a PCL file.

```
1B 45 1B 26  6C 30 4C 1B  26 6C 32 41  1B 26 6C 30 .E.&l0L.&l2A.&l0
53 1B 26 6C  30 4F 1B 26  6C 30 55 1B  26 75 33 30 S.&l0O.&l0U.&u30
30 44 1B 26  6C 30 45 1B  26 61 30 4C  1B 2A 70 31 0D.&l0E.&a0L.*p1
52 1B 2A 76  36 57 00 00  01 08 08 08  1B 2A 76 32 R.*v6W.......*v2
35 35 61 32  35 35 62 32  35 35 63 6F  69 30 61 30 55a255b255coi0a0
62 30 63 31  49 1B 26 70  36 69 36 43  1B 2A 76 36 b0c1I.&p6i6C.*v6
57 01 00 01  08 08 08 1B  2A 76 30 61  30 62 30 63 W.......*v0a0b0c
```

This is a sample of PCL and HPGL. HPGL can be recognized by the commands in capitals.

```
4C 20 45 4E  54 45 52 20  4C 41 4E 47  55 41 47 45 L ENTER LANGUAGE
20 3D 20 50  43 4C 0D 0A  1B 26 6C 31  54 1B 2A 74  = PCL...&l1T.*t
36 30 30 52  1B 26 75 36  30 30 44 1B  26 6C 30 53 600R.&u600D.&l0S
1B 26 6C 31  58 1B 26 66  31 59 1B 26  66 30 58 1B .&l1X.&f1Y.&f0X.
2A 70 30 78  30 59 1B 2A  63 36 31 32  30 78 37 39 *p0x0Y.*c6120x79
32 30 79 30  54 1B 25 31  42 53 43 30  2C 35 31 30 20y0T.%1BSC0,510
```

PostSript Print Data: Post Script is a programming language and can be complex to read. It is ASCII and can contain fonts and graphics. Like PCL, it can contain references to fonts that exist on a printer and also include font data to download to the device. Any transformation software will need to address these missing resources. Post Script files often begin with a string like "%!PS-Adobe-3.0" indicating the Post Script level.

```
25 21 50 53  2D 41 64 6F  62 65 2D 33  2E 30 0D 0A %!PS-Adobe-3.0..
25 25 54 69  74 6C 65 3A  20 50 41 47  45 31 0D 0A %%Title: PAGE1..
25 25 43 72  65 61 74 6F  72 3A 20 50  53 63 72 69 %%Creator: PScri
70 74 35 2E  64 6C 6C 20  56 65 72 73  69 6F 6E 20 pt5.dll Version
35 2E 32 2E  32 0D 0A 25  25 43 72 65  61 74 69 6F 5.2.2..%%Creatio
6E 44 61 74  65 3A 20 31  30 2F 32 36  2F 32 30 30 nDate: 10/26/200
```

PDF: PDF is Portable Document Format. Created as a viewing format it is now used commonly for print and view. In some workflows when PDF is printed there may be an interface through one of the Adobe Print Engines or products form Global Graphics that RIP (process to raster images) the file for print. PDF, like Post Script and PCL can have fonts embedded or inline and also simply referred to because they exist on the creating server's system. These are commonly called the Base 14 fonts. PDF will contain a header with the flag %PDF- and the version of PDF.

```
25 50 44 46  2D 31 2E 33  0A 25 E2 E3  CF D3 0D 0A %PDF-1.3.%......
32 20 30 20  6F 62 6A 0A  3C 3C 0A 2F  4C 65 6E 67 2 0 obj.<<./Leng
74 68 20 31  36 36 31 0A  2F 46 69 6C  74 65 72 20 th 1661./Filter
2F 46 6C 61  74 65 44 65  63 6F 64 65  0A 3E 3E 0A /FlateDecode.>>.
```

Special consideration: If the files you are processing are the product of a process that takes individual files and concatenates them to print or process more efficiently there may be many duplicate resources in these files. For example, a Post Script printer may not care if each individual document has 20 fonts inside a print file that contains 40 documents, but a transform might take every set of 20 in each document and multiply by the number of documents in the print file and create that many output fonts. This can have a dramatic effect on font numbers, file sizes and transform times. The AFP target may not want 300 fonts which can be detrimental to the transform performance. There are ways to deal with the issues if they are known ahead.

Where will the Transform Run?

If files must move between physical computing platforms there are some considerations. If the data and resources are on a mainframe and they need to be moved to a server platform there is a specific protocol to prevent distortion or corruption. As a general rule, movement from server to server, as long as it is done in a binary format is usually correct.

Xerox –Moving from Mainframe to Server

Do not translate any print data. Move the data in a binary manner. If the data is in a fixed length data format, note that length for the transformation software. If the data is a variable length dataset, there will be some sort of length field prepended to the records so the software can process in record order. The transform vendor should be able to address this. The compiled Xerox resources: Forms (FRM), fonts (FNT), logos (LGO), images (IMG), Page Descriptor Entry (PDE), Copy Modification Entry (CME) should also be downloaded binary and should not need a length field. The Xerox source file, the JSL should usually be readable on all platforms so it should be downloaded as text and readable on the server.

AFPDS –Moving from Mainframe to Server

Do not translate any print data. Move the data in a binary manner. If the data is in a fixed length data format, note that length for the transformation software. AFP files contain their own internal length fields so no additional length files are needed so the files can be downloaded binary. This is true of the resources; download them in a binary manner too. The only exception is line data with some x5a records. This is not at all common but the data may need special considerations by the vendor to download them.

Things to Consider

- Know where the data to be transformed comes from, and learn what steps it goes through before it reaches its final form currently.
- What will the data be expected to do or have done to it after the transformation?
- If it is being transformed already by another process, have matching sets of data for the new process to test with, as any new transform will be compared to the old.
- Get an overview of where various processes occur and what mechanisms are used to move data from system to system
- Understand what "resources" mean for your data, and where these resources are located.
- Moving or acquiring resources for a transform can be complex.
    - Map out the path the transformed data will take in its journey to final format.
- There can be further manipulation of the data by other processes with standards the new format will need to meet.
- File sizes or other considerations that affect storage or file movement may impact the process.

Educating yourself on you data and the processing it currently goes through will enable you to enter the transform process with more assurance of success. Learning as much as possible of the later path of that data will ensure any new process also succeeds.

Print Management

Contributed by Franklin Friedmann, edp

What is Print Management?

There are many definitions of Print Management. It is most simply defined as the management of the process used to deliver output to a marking device. Why do we need print management at all? If the printer is directly tied to the application that produces the output, only one job at a time can be processed and printed.

Since emerging as a discipline in the late 1950s, Print Management has evolved and grown in sophistication from lists on white boards to a series of clicks in an application to schedule a task. To understand the role of Print Management today it helps to understand where it began.

The Start

Initial Print Management implementations were a multi-stage affair. A limited operating system on the printers, whose sole purpose was to accept one job at a time, required an independent two-stage management process usually starting with a print file sent to an external tape drive. Print jobs were collected sequentially before the tape reel was removed and then loaded to a tape drive directly connected to the printer. Jobs printed one line at a time with no variability; whatever was on the tape, correct or not, printed.

Evolution to a Single Manager for Multiple Printers

A computer Operating System capable of handling multiple tasks and subtasks arrived in the late 1960s. A Print Manager followed on the heels of this capability, the most notable being the IBM Houston Automatic Spooling Priority (HASP), developed in conjunction with the National Aeronautics and Space Administration (NASA – Houston). It was the forerunner of the Job Entry Subsystem (JES) available on mainframes to this day. The spooling system took advantage of partitioning to operate multiple activities within a single partition, simultaneously receiving files for print, managing print selection, and printing to a device, each in a separate sub-task. The spooling tasks could include other input/output options such as card reader input and card punch output.

With an intermediate disk drive repository for the print data, use of job queues became possible. Now the job print sequence could be based on algorithms, such as priority queues and priority jobs within the queues, as well as operator selection and reorganization. Rudimentary commands allowed print job restart at a selected point, back up line spacing, and other commands. Programmable exits made job accounting possible as well as inspection of job control information.

Print start and stop, execution time information and line print counting for impact printers provided the start to user billing to accurately pass charges for print back to the requesting organization. As print batching capabilities expanded, identification of user jobs led to the addition of cover and trailer pages with at least job name, file name and time stamping. Exits permitted additional information to be appended to identify user names, departments and other information beyond simply the job name.

Now the print process could be dynamically customized. Job parameters, some relating print queues and priorities, could be modified. Exits also provided basic data manipulation as each line of data was intercepted for inspection and potential modification.

Despite these enhancements to the print management function, print was "as is"; whatever was received was printed. Job name, sender and content limited the information available to the operator. Whether the printer was operational or not could only be determined by the number of print lines remaining or more likely walking over to the printer to determine its state. Some larger print facilities added a not ready light to ease the surveillance.

When printers were remotely connected by telephone lines, a remote operator console for the purposes of starting, stopping and selecting print also appeared. The remote operator console initially used a typewriter-like device on a low-speed phone line. As the technology improved remote operators could view and manipulate jobs at the mainframe queue.

With the introduction of page printers in the 1970s, print services took on a larger role. Initially the printers appeared as impact line devices to spoolers; they accepted the lines of print set up to identify fonts, line skipping, and spacing at the printer. Some printers stored the fonts and formatting instructions within the printer activated by instructions in the print file. These printers also downloaded the print jobs before starting their printing. Others required an enhancement to the print management software to download the various resources based on instructions to the printer before each print job.

Alongside the emergence of multiple font capability came strategies to handle the font changes. Initially limited to four fonts, a character position was added to the line of data to signify which font was to be used. The column for the "carriage control" was also liable to move away from the first position of the print line. The Xerox line conditioned data stream (LCDS) to printers was one of the first such implementations in the 1970s.

At this point the character mode of operation gave way to encompass the electro-photographic technology in the printers; resolution enabled text of varying sizes and qualities, graphics and positioning beyond line spacing. Xerox Metacode and IBM Advanced Function Printer Data Stream (AFPDS) became the most common production print data streams to support this capability. This challenged spooling systems due to a larger use of memory and disk to accommodate the files, plus a new way of measuring these files, by the page (actually by the impression as in one side of the page). Restarting a job now had to take into account each page boundary.

Networked High-speed Printers

Printer spooling initially meant routing output to devices connected to the host, either in-house or at remote sites, with the digital printers controlled by the central process. Hindered by the costs of telecommunications and printers themselves, networks were limited in scope. The operators still understood the jobs as "lines of data" to be managed.

As the costs came down for printers and telecommunications, larger networks were developed. The insurance industry took a lead in remote printing in the early to mid-1980s, recognizing that the cost of the printers was in the $50K range for devices that reached 50 impressions per minute with running costs of multiple cents per impression. Many companies in the insurance industry incorporated a small print server at each location to manage the printer, receiving print files through the night on the cost effective low speed lines, with actual printing performed in the morning. Where a server wasn't included, the printers were point-to-point on dedicated lines. IP technology did not come into use until standardization in 1998.

Office Printers

With office printing requirements growing, once a printer was at the remote site it became worthwhile to share the device, delivering alternate data streams found in the office. PCL and PostScript, initially considered desktop device data streams, could now be distributed to any printer capable of receiving it. These printers could do double duty with connections to the central site and locally print jobs, as well. Initially, the print server needed instruction; a PCL and PostScript capability had to be defined for each printer, otherwise a data stream sent to an incapable printer just spewed out odd characters, often running through pages of stock with a single line on each.

The choice of papers required another level of management for the variety of paper types and post processing requirements. The print queues for cut sheet devices had to account for the number of input trays / bins along with the paper type to be used in each. To print efficiently, the spooler had to keep track of the type of paper placed in each bin, placing holds on jobs until the operator acknowledged the proper paper had been loaded. This process was not automated as no printers could identify what paper they held. Operations or systems programming had to set up print queues to minimize paper changes.

Managing Networks and Print Data Streams

The data stream also had to reflect new capabilities in printers and job spoolers. The IBM approach required a twofold upgrade. The first was a means of predefining the printer characteristics, ideally discovered from the printer. The second feature was a protocol that defined itself with a bi-directional communication between spooler and printer, interrogating the printer to determine capabilities before transmission and then verifying that the resources were in the printer or downloading them.

IBM's Intelligent Printer Data Stream (IPDS) in the early 1980s was one of the answers for production class page printers. As market requirements changed to embrace multi-channel delivery, the data stream name evolved to the Intelligent Presentation Data Stream. This capability took AFPDS and added print device specific commands that allowed the spooler to react to the needs of the printer before continuing. This ranged from resource downloads to recovery through a restart, knowing what page was last produced. This required the spooler to constantly update its understanding of page completion, receiving messages of print success as the job progressed.

Adding Personal Computers to the console was the next development as the mainframe green screen offered only lines of data. The initial implementations were built around a single server to a single printer, delivered from a single high volume mainframe. As the printers came down in cost the Personal Computer also gained dramatic performance improvements and came into its own in the 1990s, though until Internet Protocol (IP) addressing they saw limited implementation. A display, now often called a dashboard, contained the job information and more valuable color-coded printer activity for quick status evaluation by the operator. As an information center, start and stop times could be tracked and delays identified.

Queues could be defined for single or multiple printers with one queue to many devices to spread work and provide printer redundancy. The queues could relate to individual printers or like groups of printers, defining the acceptable data streams, the ability to assign print priorities.

Input from multiple sources/hosts could deposit their jobs into queues, or hot folders, though a limited amount of information came with jobs to identify their provenance as metadata. A limited solution was to have one queue / hot folder per sender.

Network managers with disparate printer types distributed across geographies found that adding new devices, addressing, and functional characteristic definitions created a major operational headache. An effective improvement came with the IP addressing scheme, where the print server could search for printers across a range of addresses to find newly added devices and define a standard device metric. This dramatically reduced network maintenance where more than a thousand printers might be the norm for an enterprise or governmental organization.

Transforms from one printing protocol to another added flexibility where printers might not be able to handle all the data streams. This required a definition of the printer capability against the understood source queue to be automated. For example, when a PostScript file arrived on a PostScript queue but only a PCL printer was available, the transform could be configured to start automatically as the job moved to the print queue. At a later point, automated data sniffing identified the original data stream, but there was no available process to recognize the printer's attributes.

Print servers tended to emphasize the functions expected in the particular environment. Print servers on UNIX, Linux would differ from OS2 or Windows-based functionality, with different file delivery commands, transforms, levels of manipulation and job acceptance. Requirements for job checkpoint and restart, job recovery or job reprint functions tended to vary, with financial institutions requiring the most elaborate functionality.

Bi-Directional Network Protocols

The next step in the evolution was the development of bi-directional printer protocols to establish a conversation between the server and the print device.

SNMP (Simple Network Management Protocol), a network protocol used to monitor and manage network devices, was developed in the early 90s. Printers, being network devices, had various MIB's (description of information about a device) which covered information from the printer's state, media loaded, and status of submitted print jobs. With this information, print management tools determine which printers are available and have the resources required to print a job. SNMP is widely used in distributed office environments where being able to remotely monitor printers is beneficial. Most production equipment is also SNMP-enabled; however other more optimized methods are typically used.

Internet Printer Protocol

An Internet Printer Protocol (IPP) was introduced for discussion in 1999 with the standards evolving through 2005. This new capability offered the ability of the network manager to interrogate the printer capability and potentially restructure the job! Consider today's laptop needs; a user in a hotel wants to send a file from the hotel room and is provided with the address of the printer. The objective is to print without initially knowing the exact printer capabilities. Using the IPP, the PC first determines if the print job can be sent "as is". If not, what alterations must be made? A particular impediment (and annoyance) is the shortage of memory when transmitting large images. One would prefer to know this before transmission. IPP, similar to IPDS, provides a method to do equipment discovery, transport the print file, provide printing instructions, and monitor print jobs status.

Job Definition Format and Job Messaging Format

In a different paradigm, JDF (Job Definition Format) and JMF (Job Messaging Format) provide similar capabilities, but are generally used to describe how a print job needs to be manufactured. It lets the destination decide the details of how it is done. JDF provides information that includes how the document is initially composed, attributes of how it should be printed, as well as how it needs to be finished, packaged, and shipped providing a start to finish description.

JMF provides the transport for the print document, device information for print management tools, and sends details to Print Management Information Systems (MIS) often used for billing and tracking. These characteristics are not seen in other protocols. JDF/JMF is used with PDF print files, which are the industry standard in graphic arts printing and less common (but growing) in transactional print. With the run lengths of transactional documents coming down in some industries, the use of JDF and PDF are becoming more viable, particularly as some traditional transactional documents are being repurposed as short run transpromotional pieces with graphic arts requirements.

Managing Print

Print Reduction Options

Through a combination of workstation and print server software, a new class of print management software debuted. It puts a barrier between the user and print using features that require authorization to print, or deny access to print services. Each access type is determined by rules that have been defined for individuals or classes of users, as well as the type of print job.

These solutions were introduced as color printers entered the network, the immediate saving came from prohibiting the use of color when it might cost ten times the cost of printing on monochrome devices. Today with ink-based solutions, the distinction is not as great. The use of queues to manage print begin to control print to more expensive devices has evolved to queue-based management for a variety of business purposes beyond simple print reduction.

Print Server as a Business Model

Technology improvements evolved and offered the option to change how the print server is viewed in the context of the business operation. With improved storage and faster processors, and the addition of Internet access, intelligent job submission emerged.

Consider the situation where organizations not only want to print current files, but also print quality brochures or forms on demand. These brochures and forms can be managed in a digital asset management or content management for the corporate client staff to access as needed (within their permission structure). Staff heading toward a business show now sign on to print management function to order the types and quantities of pamphlets that are printed on demand instead of picked form a warehouse.

Print management in this context encompasses Web-to-print tools. Now the same systems used to manage transaction documents also manage supporting marketing collateral. When integrated with inventory management the Print Management function becomes the focal point of activity.

Print Server Expands its Role for Commercial Printers

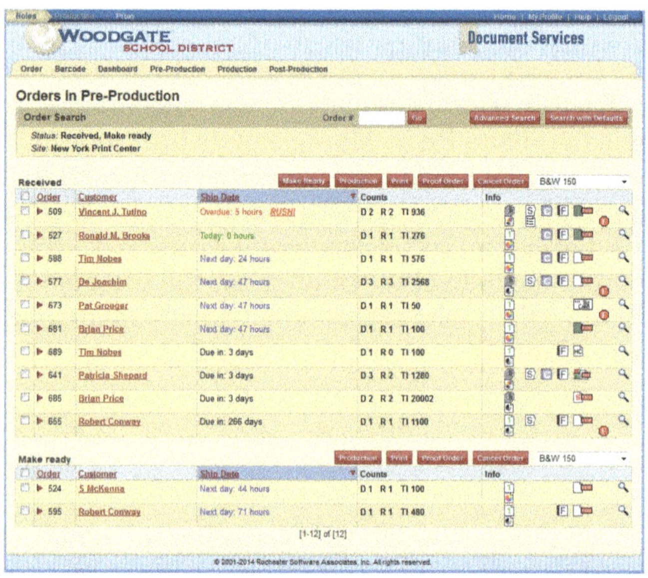

Many print server solutions have added front ends to become e-Commerce platforms. Not only as a client submission utility, but as a store front with composition, color management, job sizing and pricing, providing the functionality of a Print Suite. The client can track the job production throughout. At the back end, alternate output strategies and archiving options permit the Print Shop to have an expanded working relationship with their client.

Costing Principals

Defining the elements in a Print Process; consider yourself a printing business that must recognize all expenditures. Elements include:

- The print server itself: one-time costs for hardware & software; annual maintenance costs for the same. Include the floor space, power and cooling necessary or the monthly costs for the device if provided by another organization. May include the network construction and integration.

- Staffing percentage to manage and upgrade, assuming this is a proportion of a Full Time Equivalent (FTE) individual with the requisite skills, if part of your organization.
- Number of operators to perform the tasks. Operator training on the use of the Print Server, initially from the supplier. Refresher or training for new personnel required.

Costing for the print operation must cover:

- The printers, their initial costs, supplies, usage and maintenance,
- Print device supplies such as paper required to produce the output,
- Inserters, including maintenance (possibly parts that wear heavily such as knife edges etc) Usually requires a second set of operators with different training, and
- Postage for transactional print; possibly courier services for bulk shipments. This historically has represented 60 – 75% of the total operational expenditure.

The Evolution of Workflow: the ADF

In high-volume production print in the early to mid-1990s jobs were essentially printed as they arrived in the print queue. Then came the requirement to manage and then manipulate the data stream. Starting with print job receipt through insertion into an envelope, the entire process became the subject of detailed, automated management. Coined in 1996 by Gartner, the Automated Document Factory (ADF) identified multiple stages in the flow of data from input through distribution. These stages were identified in what is now ADF Version 1.0 and 2.0 (2007). The Workflow represented multiple possible steps that could be incorporated for given jobs based on an initial setup.

The stages are:

- **Receipt of data and transformation:** This input phase covered not only the data, but commands or instructions on data manipulation. This required a form of instruction management that would later be used against the particular file. Identification of the suitable instruction set would most often be automated through the job name.
- **Data transformation:** Initially aimed at data stream conversion (for example, AFP to PDF) it might also include using a utility that could take line data using mono pitch fonts and reformat it into a document using typographic fonts. The key is that the composition process became integrated into and managed by the ADF flow, and reported back on success or failure before the next step.
- **Content modification:** Adding graphics (pie charts, line or bar charts) that might take information existing in the print file (for example, statement data) and build relationships, such as percentage of financial holdings representing net worth. The key facet was the ability to perform this graphic addition without returning to source information, though in general, preference would be given to the composition utility performing the action and storing the information before presenting to the ADF.
- **Reformatting:** This includes reformatting address information, relocating it on the page to meet requirements for address windows across applications,

reducing the number of envelope types needed to perform the print center's work. When businesses built print applications separately, it was not unusual to have 20 or more envelope types in use. Reduction to fewer sizes and window locations reduced costs and simplified insertion verification.

- **White space management:** Finding unused space on the page in which to insert messages in a predetermined priority sequence until the space is used. This represents an extension of the application without returning to the original preparation. The choice of messages varies by individual case, and is tracked as part of the output.

Managing Inserts

Before ADFs became common there was a sticker in use in many print rooms that said: ***Sorry, Damage by Mailing Equipment***. ADF-managed print environments eliminate the problem.

The need to manage inserts added requirements for bar codes to encode the number of sheets expected in a document ensuring that the proper number of sheets were in each envelope. In some systems a sequential number on the first sheet in the envelope window for each statement is monitored by a scanner to ensure double stuffing (two statements placed in a single envelope therefore the sequential number has a skip) doesn't occur.

At the inserters there are often supplementary stations that include preprinted notices, advertising, or return envelopes. As inserter manufacturers provided their own inserter bar code definitions, which may vary by device model, the ADF is tasked with the ability to provide a replacement bar code should a reprint with insertion by a different insertion device be planned.

Managing the insertion process requires tracking each statement in the print stream so that a reproduction can be produced if there is damage to one or more pages in the document. The Best of Breed approach is to throw out all pages of the damaged statement that may have been damaged by the inserter rather than simply reprint a single page and insert by hand. The objective is to remove a primary cause of error in producing mailed documents: human intervention. This is sometimes managed by File Based Insertion, a tracking process whereby each sheet is tracked through to the envelope.

Dashboards are the preferred method to provide operational and management surveillance of the process. High level elements of all the steps are generally indicated with drill down capability against a job a particular step or a device.

As the systems become more sophisticated, Service Level Agreements have become an essential part of operations. Incorporating a particular set of characteristics of a production job, SLAs include expected arrival and the required delivery. Notification is generally required when the job will be late at any of the operational stages, job in, through mail or courier out schedules.

Billing also comes into play. The ADF is expected to calculate mailing costs per document for direct payment to the Postal Service. As postage has been calculated at 70% of the total "printing" costs, including hardware, services and staffing, this is no insignificant amount. The ADF will also prepare the information of the number of envelopes produced for mailing and in some environments send the results to the post office with payment arranged.

From a management perspective, the statement preparation log can perform double duty. It can identify every statement that has been prepared for the mail, in many countries the accepted measure of the sender's responsibility to delivery. For this reason the post to mail becomes a log of statement completion is documentation that should be maintained were it to be required for legal purposes.

Justifications for an ADF Workflow

- **Service Level Agreement:** better reporting toward on time delivery,
- **Digital management:** single level of control regardless of destination; digital delivery implementations should include simplified payment processes,
- **Cost control:** stronger relationship between work and its management,
- **Compliance:** increasing compliance requirements to respond to government legislation, including security, is leading to the implementation of effective piece tracking with return message management where possible,
- **Post:** Postal Service document production and savings, recognizing that each country offers different price rules and pricing scales; current workflows feature technology to provide document house holding, a means of combining multiple statements in a single envelope where permitted and acceptable to the recipients; sorting for the lowest cost mailings may also be permitted,
- **Continuity:** business continuity represent a means of switching from one server to another, including known job status in a time period deemed appropriate for the organization; as mentioned previously, this may be milliseconds for online systems; minutes to hours for print,
- **Recovery:** the recovery commitment will determine the architecture and resulting cost structure. Secondary servers can be "hot standby", print server application startup, or alternate location management with the existing printers; the first two require common print spool storage accessible by each server; pricing is usually available for all secondary server running options, and
- **Workflow Extension:** extending the Workflow upward to include "document capture and conversion", a scanning and storing process is part of the smaller general purpose workflows that are being implemented to encompass the entire realm of paper ingestion through print / archive.

Robustness; Security Considerations

We expect that any system to manage critical operational information be robust. Sending out statements must be recoverable as discussed in Disaster Recovery.

Equally, backup for print centers and the Document Factory servers must be provided. The design of these ADFs can include offloading to alternate sites, external printer management with secondary servers and additional functionality, supporting a Disaster Recovery strategy.

An interesting approach taken since the early 2000s is the implementation of an overload provision to in-house systems. Rather than buy additional printer for peaks, some jobs may be outsourced, ADF to ADF, for completion. This has limited applicability if unique paper requirements (i.e. logoed paper) are not stocked at the alternate location, though it has become easier with color digital printers able to work with white paper. This process still requires a substantial investment to develop the revised print application and the replacement electronic forms.

Security is essential therefore passwords and permissions must be an integral part of the process. Control levels to limit the breadth of action by an operator are required parts of any design. While deleting a print job might be appropriate, you would not want to allow an operator to deliberately or inadvertently delete an active printer definition from the spooler's available devices. When it happens it requires immediate action by an individual who can reconstitute the lost printer AND its definitions. Depending on how the print queues had been defined, this may require significant recovery activity. A comment on the need for security against malicious acts; print may be delivered to the spooler from external sources. It is essential that the organization is protected.

Hosting: In-house or External

Implementations have improved to provide two sets of options for print servers:

- In-house print servers, the process that has been available in various guises to date.
- Externally web-hosted servers with the same functionality, with the financial justification developed on removing the print application maintenance from the client's already stretched support staff while operations are unchanged.

Where to Next?

Technology has brought changes to the world of print. Many ADFs act as a platform for output, repurposing documents for electronic distribution. This will only increase. To meet this evolving output usage will require new models:

- A "pull" model, requiring action on the part of the recipient. This is accomplished today with QR codes.
- A "push" model, sending out the electronic equivalent of Direct Marketing, requires distribution addressing. Today this is managed when Internet subscribers signup for a service.

However, while the last step of output tracks each mail piece or message sent, proof of delivery is another matter. In mailing, registration with the delivery organization serves this purpose, email or file transfers usually have a return request blocked by the recipient's server, a procedure developed to avoid unscrupulous use of the Internet.

Nor does regular mail currently provide a reliable means of tracking delivery in all countries. The Postal Services endeavor to deliver to the identified recipients and may in the case of error return the mailing, usually for a fee. The list of recipients for whom mailings have been produced and given to the Postal Service is a demonstration of logging controls and may be a sufficient equivalent to unregistered mail delivery.

Further automation should also be expected. From reprint processes through Business Continuity, output management will be simplified for operations staff.

Security, auditing, and compliance mentioned earlier will receive greater emphasis as organizations face an evolving business environment. These capabilities will be either standard or optionally packaged. Organizations most heavily monitored (the financial and health industries) will lead implementations, as they have in the past.

How will print integrate with "socially acceptable media"? As social media demands immediacy and greater relevance than a typical mailed statement, there will be a need to integrate new sources of information. While this would be the purview of a composition utility, the logical placement of this action will need to be close to the message distribution to make it timely.

References

Print Services Facility for z/OS Customization S544-5622 http://publib.boulder.ibm.com/infocenter/zos/v1r13/index.jsp

CA-View Output Archival and Viewing System Reference Guide (formerly Unicentre) http://www.ca.com/us/products/detail/CA-Spool.aspx

IBM eServer iSeries Printing VII: InfoPrint Server Implementation ibm.com/redbooks http://publib-b.boulder.ibm.com/Redbooks.nsf/redbooks/

Levi Ray and Shoup Inc, Spool products http://www.lrs.com/eom/PDF/V2R11Products.pdf

z/OS Infoprint Server Customization S544-5744 http://publib.boulder.ibm.com/infocenter/zos/v1r13/index.jsp

IBM IP Printway Guide S544-5379 http://publib.boulder.ibm.com/infocenter/zos/v1r13/index.jsp

StreamServe Business Communications Platform

RSA WebCRD Integrated Workflow http://www.rocsoft.com/

Storefront at EFI www.efi.com

IBM z/OS JES2.... http://publib.boulder.ibm.com/infocenter/zos/v1r13/index.jsp

Solimar Enterprise Output Management http://content.solimarsystems.com/pdf/SPDE_Benefits.pdf

Pitney Bowes DF Works http://www.pb.com/software/Print-and-Mail-Production/Automated-Document-Factory/

Ricoh PrintProcessDirector http://www.infoprint.com/internet/ipww.nsf/vwwebpublished/solos_automated-document-factory_en

What is JDF? http://www.cip4.org/overview/what_is_jdf.html

Electronic Presentation

Contributed by William Broddy m-edp

What is Electronic Presentation?

Electronic presentation covers the many ways a document may be presented to a customer on their computer screen, on their phone, on their tablet, or even on their TV screen. As the delivery of a transaction document moves from physical production and distribution to electronic presentation it is critical that it meet the same business, contractual, legislative and evidentiary requirements of the analogous print and mail process.

The documents we physically create, package and distribute must be:

- **Legally presented:** there must be a chain of custody to the other party, or to a trusted agent,
- **Irrefutable:** they must prove who issued them and when,
- **Permanent:** they must be created in a way that avoids surreptitiously alteration,
- **Accurate:** the information must be authentic, and
- **Usable by the recipient:** they must have no encumbrances or dependencies from the issuer.

Proof of Receipt

The mail box, post office or postal rule is a part of the legal rules of evidence in most countries. Courts presume that a document placed into the postal system will be presented to recipient[1] . The issuer of the document only has to provide proof of a chain of custody from the beginning of the creation process through to the posting of the document at the post office. More recent citations require proof that the posted document has the correct address and adequate postage. If these criteria are met, it is presumed that the document has been delivered and the onus is on the recipient to prove that it wasn't.

Under the Accepted Principles of the Model Law[2], an electronic document is considered delivered "when it enters an information system outside the control of the originator or of the person who sent (it) on behalf of the originator". If the document is parked on the issuer or issuer agent system, the document is not considered delivered until "the time when (it) is retrieved by the addressee."

---

1 Example of Mail Box Rule being cited in court rulings: No. 10-5205. United States Court of Appeals, Sixth Circuit - SOLOMON OLIVER, Jr., Chief District Judge Filed: October 6, 2011 - "For the presumption of receipt to arise under the common law mailbox rule, a party must present evidence that the letter was "properly addressed, had sufficient postage, and was deposited in the mail". WIGMORE ON EVIDENCE states: "The presumption [of receipt] rests upon the supposed uniform efficiency of the postal service in delivering letters duly stamped, addressed, and mailed into its custody; if therefore the efficiency is operating, does not the non-arrival of an alleged letter indicate that such a letter was never given into the postal custody?"

2 UNCITRAL Model Law on Electronic Commerce. http://www.uncitral.org/uncitral/en/uncitral_texts/electronic_commerce/1996Model.html

Postal authorities have a unique position in this environment. Common law in most jurisdictions presumes that a document presented to the post office is, in effect, delivered to the recipient. Any electronic document presented to the post office (electronically) to be delivered by them to recipient would be considered presented once in the post office's possession[3].

Portability

Can the recipient freely read and use the document in the future? Can it be forwarded to another party intact? Can the rule of acceptance be met? A legally presented document must be completely under the control of the recipient. It cannot refer to information that is only available within the issuer's environment. For example, a bank can't tell you that you can only see you current monthly statement at the local bank branch and can't remove it.

Electronic documents that are presented must have the same portability. The recipient's document must contain all document objects necessary for faithful rendering, at the time of viewing and afterwards when sent to interested third parties. The portability applies to both the information and layout of the document, and the electronic proof of authenticity, such as a digital signature.

Permanence

Can the document's content, perceivable look or proof of origin be altered? One facet of a paper transaction document is its permanence. It is very hard to alter an electro photographic or ink jet document. The document carries the look and feel that the issuer requires and will contain nuances that are hard to counterfeit. Interested third parties accept it as legitimate: courts accept it as evidence, banks as proof of net worth, accountants and taxation authorities as proof of an expense.

Unfortunately, most electronic communications methods can't provide this level of permanence. For example, HTML pages are interpreted by the recipient's web browser. The layout cannot be assured, and critical information may not be displayed. It is very difficult to capture and archive an HTML page in a way that it would be acceptable to many interested third parties, especially if reproduced in the future.

Authenticity

Does the document's metadata (digital signature) prove who issued the document and when? Was it applied during the normal course of business? Paper documents provide a date of issue, the issuers' logo and other look and feel attributes. It is delivered by the post office in an envelope that identifies the issuer (hopefully with branding, but at least with an indicia mark).

An electronic document must contain similar proof of its authenticity. The Model Law intends to "allow for the authentication of (a document) by means of a (electronic) signature[4]" and ensure "that a document would be in a form acceptable to public authorities and courts."

---

3 Canada Post's ePost offering is an example of this. http://www.epost.ca
4 UNCITRAL Model Law on Electronic Commerce, Section 47 – Point 6

Legibility

Can the document be read or perceived by the recipient as the issuer intended? A printed transaction document is laid out as the originator intended. The composition process ensures that the document layout remains consistent and that the recipient receives a document that is legible and understandable.

One of the intents of Model Law is to ensure in electronic commerce "that a document would be legible by all"[5], and "that a document would remain unaltered over time and provide a permanent record of a transaction "[6].

In our industry the requirement is to adhere to the above requirements to ensure the electronic documents presented are acceptable to the courts and other interested parties.

Technology Requirements

There are four components to electronic presentation: the format of the document being presented, the process by which the document is created, the mechanism to either deliver the document or a notice to get the document, and the process that ensures that presentation took place.

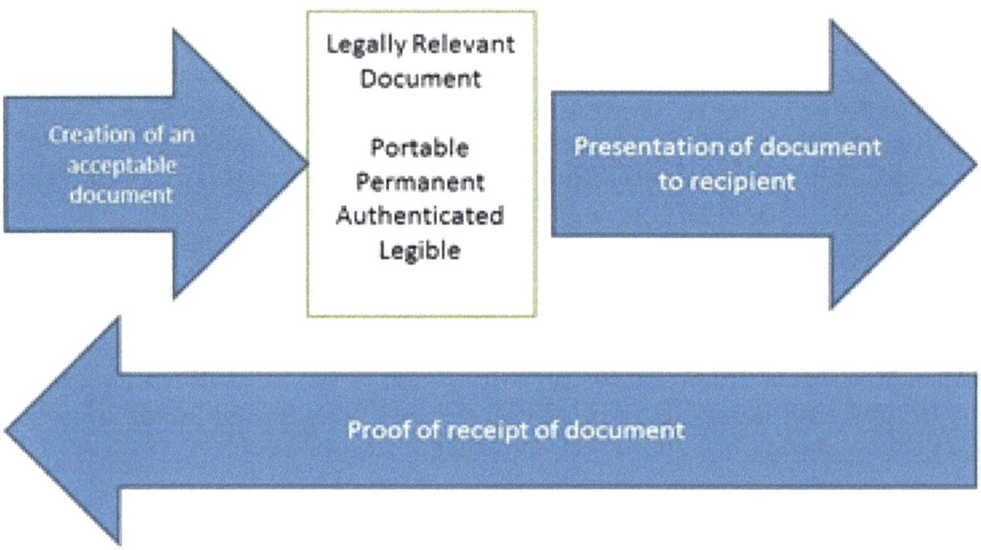

Figure 1 - Electronic Presentation Workflow.

Formats

The most important technology to decide upon is the document format. It must meet the requirements of the Model Law.

5 UNCITRAL Model Law on Electronic Commerce, Section 47 – Point 4
6 Ibid Point 7

Bitmap Images

BMP, TIFF, and JPG file formats have historically been used for easy delivery across different platforms when there is a requirement to store a record of the document exactly "dot for dot", i.e. the electronic version viewed on a computer monitor should be an exact match to the document created on the printer.

This method has gone out of favor as there is no universal standard on what bitmap image format is the best to use and therefore no standard on reading software. Bitmap images of individual pages means that they are typically large in file size when compared to other methods of storage and delivery. This format also lacks the ability to contain metadata in a standard way within the file itself.

However, bitmapped images digitally signed with proof of originator and time of transmission, meet most of the criteria laid out in the Model Law.

Object-oriented Documents

The most common formats delivering final-form documents are variants of Adobe Portable Document Format (PDF), although presentment solutions do exist for AFP.

PDF

PDF, governed by ISO standard 32000-1[7], has been widely accepted as a format for presenting final form documents. However, there are many flavors of PDF that are provided by composition or transform utility suppliers. At the lowest level, there are PDF files that contain nothing more than a TIFF image for each page. Although viewable, they are large and have no search capability. At the high end, there are PDFs that contains objects that may require special extensions or viewers. Although PDF files do provide a final format they can be altered unless properly locked.

A properly structured PDF that is digitally signed through a credible process may meet the evidentiary requirements of the Model Law. However, the onus is on the issuer and software provider to prove the authenticity of any questionable document.

PDF/A(rchive)

The need for permanent, legible, authenticated documents is a requirement in most business and government environments. Corporate records management systems must meet these standards. The Food and Drug Administration requires so much original documentation that the paper evidence for a drug submission used to fill two semi-trailers. Because records managers, regulatory agencies, and government archives need an electronic alternative, they were instrumental in driving much of the wording in the Model Law and in defining the business requirements for PDF/A.

PDF/A is an ISO-standardized version of PDF designed for the digital preservation of electronic documents. The latest standard is PDF/A-3[8].

PDF/A differs from PDF by omitting features ill-suited to long-term archiving, such as font linking (as opposed to font embedding). The intent is to ensure that a PDF/A document can be opened 100 years from now and still be legible.

7 ISO 32000-1: 2008. Document management -- Portable document format -- Part 1: PDF 1.7

8. ISO 19005-3:2012 and is based on ISO 32000-1 – PDF 1.7. There is also PDF/A-1 (2005) and PDF/A-2 (2011)

The ISO requirements for PDF/A file viewers include color management guidelines, support for embedded fonts, and a user interface for reading embedded annotations.

A properly implemented process that creates PDF/A documents in the normal course of business should meet all the requirements of the Model Law. It has been adopted by the government in the United Kingdom[9], the US Federal courts, the US Food and Drug Administration, and the US National Archive among other large regulatory organizations. PDF/A is evolving into a presumptive standard for authenticating evidentiary documents, putting the onus on the disputing party to prove that the document is fraudulent[10].

Semi-formatted Presentation Formats

HyperText Markup Language

HTML is a web-based file structure that depends on the recipient's web-browser software to lay out the document. As originally implemented[11] it uses tagged logical commands to decide which font / size / weight to use. The file is *marked up* with tags to allow the web-browser to adjust the layout based on the display technology and user preferences. The heading <h1> in Figure 2 does not say where the line of text will be placed, what font / size / weight / color to use. This is left up to the browser's settings.

The information presented may be perceivably different from one browser (Internet Explorer) to another (Chrome) and one technology (Windows) to another (Android, IOS or BB10). The layout is not permanent.

Also, web-based information is not portable; many of the relevant resource objects are not contained within the file, but linked at the issuer's website. If the recipient saves the document "for subsequent reference", these resource objects may not be available if the recipient tries to look at the document offline, the issuer removes the objects from their website, or blocks the recipient, or interested third party, from retrieving the objects (e.g. the recipient no longer has an active account, or the interested third party cannot log into the recipient's account).

```
 <!DOCTYPE html>
<html>
  <head>
    <title>This is a title</title>
  </head>
  <body>
    <h1>So Long</h1>
    <p>and thanks for all the fish</p>
  </body>
</html>
```

Figure 2 - Example of a marked up (early) HTML file.

9 For static versions of non-statistical data produced for download, archiving and authenticity, PDF/A should be used as the default for non-editable documents, http://standards.data.gov.uk/proposal/viewing-government-documents

10 This should not be considered a legal opinion.

11 RFC 1866 - Hypertext Markup Language - 2.0 I

The HTML files do not contain a digital signature that would authenticate the issuer, time of issue, and integrity of the contents. This information is carried in the transmission layer (S/HTTP) and does not stay with the file once it is saved.

Finally, browser recipient settings may cause the information in HTML file to display in a way that is impossible to understand (e.g. browser accessibility feature may remove background (CSS?) layer that provides key static information or changes all shaded fonts to 100% black). This may cause the document to be illegible.

eXtensible Markup Language

XML has a tag structure similar to HTML except that it only provides a way to define tags for specific environments. It provides the syntax that various XML Schema (something like a profile) must follow in creating new tags. There are hundreds of XML Schema in use, many of which are transparent to us. In most cases, the tags within the Schema are used to logically define data, not place it. To use XML-defined data requires the use of an eXtensible Style Language (XSL) file. To use an XML file to present a legally relevant document has most of the portability and authenticity limitations of HTML

Creating the Electronic Document

The two most common ways to create an electronic document are to compose it or to transform an existing print file into PDF.

Composition

Most current composition products used to generate transaction documents offer PDF, PDF/A, HTML, XML, and TIFF emitters. If you have one of these products, it is likely that you can generate PDF and PDF/A that are perceptively identical to their printed counterparts.

You can also generate HTML and XML that can be fed to an external system that displays the file. The end result might even look like what you intended. It will not, however, have the look and feel of the printed equivalent circulating in the marketplace. It will also have all the limitations identified above.

Transformation

Many transaction document applications run on legacy platforms. Still others use COBOL, C and even PASCAL programming to generate documents. The most practical way to create electronic documents from these applications is to transform the existing print file into the desired format.

There are many excellent transaction document print stream transformation tools that do this. You should ensure that they create more than images of each page in the file.

Presenting the Electronic Document

There are a few ways to deliver legally relevant electronic documents. If you have created a document that is both portable and has proof of authenticity, you must present it in a way that ensures timely delivery, and can be proven.

Internet Email

Although sending email correspondence is well established as a reasonable way to communicate, there are some inherent limitations around proof of delivery.

The most common method of sending email is using Simple Mail Transmission Protocol (SMTP). This allows an email object to be submitted to the Internet where it will eventually be presented to the recipient's email server. Along the way, a copy of the email is temporarily left on every server it touches.

Once it gets to the email server, the email may be considered spam and quarantined or erased, or it may be approved and held for the recipient to retrieve, using either POP3 or IMAP.

Unfortunately SMTP provides no chain of evidence back to the issuer, only to the recipient. Although a read receipt is possible the issuer is dependent on the recipient to allow the response to the request. Email formatted as HTML may not actually reach the recipient's email system.

HyperText Transmission Protocol Secure (HTTPS)

Web browsers use HTTP to call a file for viewing. The browser must use an HTTPS call when the webpage contains or requests sensitive information. HTTPS provides authentication of the webpage owner (through a third party-issued digital signature) and provides bi-directional encryption between the browser and the server.

HTTPS can be used to call HTML, PDF, and XML files.

HTTPS is a passive process. It requires the recipient to proactively fetch the document from the website. The recipient may not visit their mailbox frequently. This affects the recognized time of presentation, depending on whether the document is parked on the issuer's server, or on a third party server that is arms-length[12] from the issuer. For the former, the time is when it is retrieved, for the latter it is the time that it is presented to the third party.

Invited Pull

The most popular way to present electronic mail is to put the document into a web-based mail-box, then send an email to the recipient inviting them to "pick up their new mail". You should not include an HTTP link in the email as this would open the process to spoofing[13] .

Mobile Apps

The third option is to use a mobile application that constantly looks for mail at third party server and pulls it onto the recipient's appliance.

This type of app would use strong encryption, and would immediately receive the document on behalf of the recipient. The app would hopefully also properly store the received documents in a way that provides subsequent use.

12 The key test of this is whether the issuer can request already delivered documents to be deleted and replaced. One you mail a letter (with a correct address and adequate postage, it cannot be returned).

13 Someone sending out an identical email with a link to a 'bogus' webpage. They would capture the sign-in information, and then pass the victim to the correct page. The villain then has all of the victim's sign-in information and can peruse their financial correspondence looking for identity information.

Proof of Receipt

A legally relevant document must not only be presented, but there must be proof of presentation. The mail box rule makes this much easier for presentation of paper documents; you only have to prove they all got to the post office. However, this is not the case for documents sent directly from the issuer. It's still unclear whether third party providers[14] will be recognized as an agent of the recipient (who gets the mailbox for free) or the issuer (who pays for delivery).

With SMTP, there is a high risk that the document will never arrive, and there would be no chain-of-custody available to the issuer. Proof of receipt is very risky.

With HTTPS, there is a clear chain-of-custody from point at which the recipient calls the document. However, there is an indeterminate latency of receipt waiting for the recipient's call.

With Mobile (and PC) Apps, there is a clear chain-of-custody and there is no latency. It will be interesting to see applications in this area mature into robust consumer receive and archive management systems.

14 One exception is Canada Post's ePost mailbox system, which is considered the equivalent of Canada Post physical mailboxes. This may also apply to other postal authorities' electronic mail box offerings.

Web and Mobile Delivery

Contributed by Neal Gottsacker

What is Web and Mobile Delivery?

Web and mobile delivery includes the actual presentation of documents on smartphones, tablets, laptops, desktops and other web-connected devices, as well as all of the document design and preparation required to support these devices. Web and mobile device display environments shape a common set of presentation options and requirements. There are also unique considerations based on device form factors and device types.

Web and mobile delivery requires consideration of the following topics:

- information delivery options,
- desired user experience,
- information management, and
- system integration.

Information Delivery

Documents can be delivered to web-connected and mobile devices in a format similar to the printed equivalent. The document may be enabled with additional capabilities based on the delivery method and receiving devices. It may even be delivered in an alternative format more appropriate to the device characteristics. When documents are delivered to mobile devices, content providers may also provide new information presentation options since the content can be organized and presented specifically for the screen sizes and capabilities of the receiving devices.

The three common delivery approaches include electronic documents, interactive documents, and mobile experiences.

Electronic Document Delivery

Some communications must be delivered as a document to meet regulatory and legal requirements. When it is permitted to deliver those documents electronically the most commonly accepted formats are PDF and image, although there are proprietary formats and associated viewers. Customers retrieve electronic documents in different ways depending on receiving devices. Web-based electronic presentment solutions generally deliver access to the document through portals, online systems, or email solutions. The documents may be handled as attachments to be opened online or downloaded from email-based links or through links found when logging on to a web portal.

Modern browsers, including Internet Explorer, Chrome, Safari and Firefox, include a PDF viewer as part of the browser. Browser-embedded PDF viewers can be disabled to allow alternatives to be used, such as the Adobe Reader, or the wide variety of freeware viewers such as Ghostview or xpdf, to render PDF documents.

Confirmation of receipt may be a business requirement for electronic documents. Since the document itself cannot communicate back to the host system without extraordinary programming it is important to identify receipt-tracking requirements and ensure that the portal, delivery system and tracking system can meet the requirements. Some portal and online systems may offer the option to record events such as document retrieval or even document open.

Interactive Document Delivery

Some documents lend themselves to an interactive experience for the customer. A variety of document types from transaction to correspondence to forms may be addressed as interactive documents. Interactive documents may be delivered as traditional electronic documents through portals, email, or part of an overarching application.

For transaction documents, interactions may allow a customer to alter information presentation using options such as expanding or collapsing transaction detail, sorting transaction tables, and displaying data in charts that can dynamically change based on user selection. Interactive documents may include hypertext links to provide navigation to sections within a document or to transfer to view external web content for associated information such as educational or marketing content.

Mobile Experience Delivery

The rapid adoption of mobile devices, including smartphones and tablets, introduces new business implications for document delivery. The form factor, or dimensional characteristics, of mobile devices and the expected user experience drive many mobile delivery design requirements.

Delivery to mobile devices is often a result of an end user clicking on a link, sometimes embedded within a mobile application, to request the document. Retrieval may be prefaced by a push notification from an application to a given users' mobile device.

Mobile applications generally take one of three forms: HTML, native application, or a hybrid approach. HTML (hypertext markup language) experiences are initiated and rendered through the mobile browser. With the adoption of HTML5, a standard that modern browsers have adopted, HTML-based experiences tend to work similarly across the variety of browsers. Since the HTML5 specification continues to evolve, special care should be taken to ensure that the expected experience is provided. Similar to websites, HTML-based applications may include live links to content residing on web servers and the ability to download files for offline viewing.

Mobile native apps are written for a specific mobile operating system, such as iOS or Android. Historically, the benefit of the specific operating system app approach was to provide app access to device features such as geolocation or a camera. However, the evolution of HTML5 has provided access to many device features eliminating many technical requirements to create native apps. A consideration for creating native apps is that the various operating systems require different programming languages to be used. For example, iOS devices support native apps are written in ObjectiveC, while Android supports Java. Cross-platform mobile application development toolkits and products are becoming available to address this development requirement, meaning the toolkit translates the app code into the device-specific language. Native apps are downloaded from an app store.

Hybrid apps combine both HTML and native app capabilities by providing a native app wrapper around the HTML application. This approach allows apps to be distributed via app stores and provides some device feature access. Technology solutions such as Adobe PhoneGap or the open source equivalent, Apache Cordova, are used to create the wrapper needed to enable an HTML app to be downloaded.

User Experience

Business drivers for document delivery to the web and mobile may be distilled down to digitizing core business processes such as billing, account notices, or marketing content for the target demographic of the electronic experience. Web and mobile platforms, on the one hand, allow for information presentation and interaction, but on the other hand introduce a new set of requirements based on the physical aspects and capabilities of user devices.

Information Presentation

Mobile delivery provides options for information presentation to customers. PDF, image formats, or HTML versions of documents designed to look like their printed equivalent require customers to scroll vertically and horizontally, depending on the viewing device dimensions and/or zoom factor. Given screen dimensions, viewing electronic documents on a PC may be an acceptable experience since pages are viewed by simply scrolling up or down. However, a document displayed on a smartphone or small tablet may require extensive scrolling or user interaction to "pinch and zoom" the screen to effectively view the information.

There are several options to consider regarding document information presentment on a mobile device. Requiring users to scroll between pages or click on links to navigate to different sections is one option. A menu bar or hypertext links within a document can be used to reference different document sections or pages or even external content.

Presentation of transaction detail also comes with choices when delivered to mobile devices. Transactional tables can be offered as sortable by clicking on a column header. Detail rows of a table may be expandable or collapsible to show or hide information. Likewise, interactive charts, such as pie, bar, or line charts, may be created to enable a user to click on a chart element to see the details. Corresponding table contents can be linked to the chart selections if desired.

Marketing information can become more dynamic when documents are delivered to mobile devices. Images may be embedded into the application or referenced as a resource through a URL. Animation through GIF (Graphics Interchange Format) files that are part of a document may be used to draw attention to a particular item. Additionally, hypertext links can be provided to reference marketing content including audio or video files.

Form Factor Considerations

A mobile device's display dimensions influence an appropriate mobile experience. For example, transaction information presented in table format that includes multiple columns may be appropriate for a PC or tablet; however, impractical for a smartphone display. Both tablet and smartphones allow a user-driven portrait versus landscape view by simply rotating the device. The horizontal space available in each mode will likely bring additional design considerations to mind.

In the case of a smartphone, transaction information may need to be presented as a scrollable list represented by one or two of the most logical table columns. For instance, the screen may allow for a transaction date and amount to be displayed with an option to view additional transactional information. One design consideration then may allow a user to select an entry from the list to navigate to a new screen that provides the additional transaction detail. Given the form factor, tablet or smartphone delivery may also require design considerations for organizing information access via menu items, tabs, links, or swiping up or across to access additional layers of information.

Mobile experiences that automatically account for the device and form factor and present information most appropriately for the particular device are considered responsive. Responsive design aims to provide an optimal viewing experience by limiting scrolling, panning, or resizing. One method to support responsive design is to establish layout and content in relative units like percentages (of the screen size). Detecting the device, or media, through queries allows the application of different design styles established by Cascading Style Sheets (CSS).

Cascading Style Sheets (CSS) use a style sheet language to describe web content presentation. Many software tools exist to generate or customize CSS for web and mobile presentment.

Mobile devices provide other features such as placing phone calls, using geolocation to determine or record location, and using cameras to create and store pictures. Some business and consumer mobile experiences may require this type of feature access to be naturally integrated. In the past, these features required a native application or a hybrid application; however, HTML has advanced to the point that pure browser-based apps written with JavaScript can also support many such device features.

Rich, interactive experiences are provided today through HTML5 with JavaScript or other technologies such as Adobe Flash or Apache Flex. With an ever-changing landscape of support, a developer should examine support for the technology used for the experience as it relates to the targeted device types. For example, currently Flash is not supported on iOS devices.

Information Management

Web and mobile experiences bring a unique dimension to application design considerations with regard to content management. Physical documents ultimately incorporate content into the document itself at composition time. Web-based documents allow for content to be retrieved and incorporated at the time of display. Since the rendering of information, often with sensitive and personal data, is performed by a distributed device, care is required with respect to where information is stored and how applications support secure experiences.

Data and Content Integration

Similar to printed transaction documents, mobile applications that deliver personalized information must account for retrieval, integration, and management of data and content. Mobile delivery does not require data or content to be stored in different locations or formats than traditional document creation solutions require. However, content such as ads, static images, or streaming information may be placed in a Content Delivery Network (CDN) to provide optimal performance. CDNs are specialized data centers strategically placed on the Internet to provide high availability and performance for certain content delivery, relieving an organization's own infrastructure demands.

Content may also be stored in Enterprise Content Management (ECM) systems for control, compliance, and brand purposes. Access to such content may require use of web services, specialized Application Programming Interfaces (API), or HTML *Get* or *Post* commands.

As with content, company and personal data may be retrieved using web services, specific APIs, or HTML *Get* or *Post* commands. Additionally, data may be housed in databases or file systems on a server. Consider the application design with respect to client versus server-side connection to databases. For example, a design including a client-side connection to a database may create issues depending on number of concurrent connections and duration connections are held open. A design practice to overcome such an issue is to manage database connections solely on the server.

Mobile Data Storage

Personalized experiences, whether as a representation of statement information or targeted marketing, will likely require data to be stored on the client device. In addition to the data retrieved from a server, data may also include user-entered information such as preferences for updates or communications.

Regardless of what information or how data is retrieved, consider how and where data is stored locally on the mobile device. Modern mobile devices and operating systems provide for local storage on the device. Some mobile operating systems include local, lightweight databases to provide application services as well. If local storage is used for an application, personal identifiable information should be encrypted. Design of mobile applications that run in offline mode, or when not connected to the Internet, should consider if and how sensitive data is stored.

Mobile and web applications may capture and store data as unintended artifacts of a session. For example, cookies, web history, browser cache, debug or log files, and mobile databases should be considered with respect to unintentional data storage. Establishing operating system settings or *no-store* directives in packet headers may prevent HTTP and HTTPS traffic caching. Also, encryption methods applied to locally-stored data are commonly recommended to ensure private data is not exposed to other applications.

HTTP *Get* and *Post* commands are methods to retrieve and send data from a client to a server. A *Get* request may include parameters within a query string. This technique should be carefully considered for risks of exposing sensitive information such as account numbers, customer names, or passwords as part of a parameter. HTTP *Post* essentially instructs a server to accept the data enclosed in the request's body. The data may be encrypted and thus helps to mitigate the security risks potentially found with an HTTP *Get* command.

Security and Authentication

In the case of mobile and web delivery, authentication is clearly required to ensure that the right user is accessing information. User identification, which is a prerequisite to authentication, is necessary to retrieve any specific data to be presented to the given user. User authentication is the responsibility of the mobile and web delivery applications.

Therefore, in addition to considerations for data treatment, mobile and web delivery solutions often require integration with corporate security and user identification systems through Single Sign-On (SSO) methods or authentication services such as the open standard for authorization (OAuth) or Lightweight Directory Access Protocol (LDAP). SSO essentially allows an application to be launched and the user authenticated by the application because he/she is gaining access through another application that has already authenticated them. SSO relies on shared authentication servers to identify users across the set of applications.

OAuth specifies a process that allows for access without sharing of credentials. It was designed for use with HTTP and relies on the exchange of tokens that provide access to protected resources hosted by a server.

System Integration

Customer communications, whether paper or digital, depend on integration with many other systems and processes. One fundamental design question that directly influences system integration requirements involves whether or not the documents or information need to be generated as part of a systematic process or as part of a user's experience where information is dynamically brought together based on the target device.

Document Composition

Mobile and web delivery solutions may be initiated as part of a document composition process. The system integration required for traditional paper and electronic documents relative to upstream data and content access can typically be leveraged for both approaches. The actual delivery could be through a repository that serves up an electronic document per user request or as an attachment or link in an email.

Inserts may be simulated in electronic documents as additional PDF documents or via links embedded within the primary communication. Since electronic documents do not benefit from postal savings or require coded values to drive post-processing equipment, barcoding is typically not required. However, some traditional post-processing steps such as report generation may still be required to reflect what customer documents were created, for example.

The document composition approach may be used to produce electronic documents for web and mobile delivery in batch or in an on-demand manner. Batch implementations often store documents in PDF or HTML format in a repository for later retrieval or send documents to users via email. Typically, on-demand generation will result in a PDF or HTML document that will be pulled by the user. In this case, the experience may be initiated from a website or portal.

Portal

Whether generated in batch or as a function of an on-demand request, mobile and web-delivered information may be retrieved through a website or portal. Online delivery sites may allow the user to set preferences for methods and events for reminders and/or alerts. For instance, a credit card customer may ask for a notice that a bill payment is coming due or that a foreign transaction was placed on their card. Notifications and reminders, while often sent from a site such as a portal, may be delivered as a text message or an email. Both methods allow for a link to be used for navigation to bill details in this example.

Portals or personalized, authenticated websites provide a landing, or home, page to enable the user to click on images or links to retrieve individual documents. On-demand implementations would mean that the document(s) are generated as they are requested. As in traditional document composition implementations, the personal data will be retrieved, the document composed, and then presented in a frame of the portal.

Implementations that use a portal approach may be extended to facilitate tracking of document retrieval and viewing.

Summary

There are several approaches that can be taken to present documents on web and mobile devices. Consideration for regulatory requirements as well as business requirements for presentation layout, supported mobile devices, desired user interaction capabilities, and system integration will influence applicable approaches for mobile and web delivery.

Retention of information in a document construct similar to a printed document tends to require larger screens found on PCs, laptops, and larger tablet devices. Electronic and interactive documents often allow users to navigate and choose how some information is displayed. Smaller screens found on some tablets and smartphones may best benefit from an HTML application or phone app that considers the form factor.

Information management and security can lead to significant technical requirements when delivering information to mobile devices. Technical design decisions related to data storage and handling of sensitive data can be particularly complex for mobile applications.

Regardless of implementation approach taken, one should realize that many new software solutions are available and continually being brought to market to more easily create electronic and interactive documents as well as mobile information presentation apps.

Archive

Contributed by Franklin Friedmann edp

What is an Archive?

Gartner has summarized the general archival process as "A consolidated system for storage, access, management and viewing of data that is often print-stream-originated. Leading uses of IDARS (Image and Document Archive and Retrieval Systems) include mission-critical customer service support, electronic bill presentment, management and distribution of report data (e.g., mainframe output, transaction logs and financial reports) and long-term archiving of historical data."

The 20th Century saw the emergence of a number of technologies based on Document Imaging. Film was used to store the pages using Computer Output to Microfilm (COM) and later Microfiche. COM technology was less expensive than paper, was eventually legally accepted in place of original paper documents. Indexes usually referred to the image location such as distance down the roll of microfilm or the card number of the microfiche. A further extension placed the information on a laser disk (COLD).

In the 21st Century the field is crowded enough to require us to differentiate types of archives based on usage. Now it moves beyond the deliberate preservation of records and remote storage to preserving with the stated objective to protect against loss. Today there are multiple approaches starting with occasional retrieval and moving along the gamut to the constant access found in Electronic Content Management systems.

Backup versus Archive

There is a difference between Backup and Archive. Where a Backup is a secondary copy typically used in recovery operations, an archived copy is intended as a duplicate maintained for analysis, value generation or compliance. Archived copies typically include context and meet regulatory requirements.

Dynamics Driving Archive Storage Management

What are the parameters that demand an archive?

- With the rapid adoption of Internet technology, the expected rate of information expansion, based on volume and the new data types taking significant amounts of storage is expected to double every 6 – 12 months. Each organization must prepare for this inundation.
- There is a staggering cost to be paid to provide this storage if everything is equally accessible. Decisions will be made to tier accessibility (also known as a storage hierarchy, where content is migrated to lower cost media as it ages and access frequency typically declines).
- To protect both users and organizations, legislation has been passed in most countries, requiring compliance, particularly in the medical and financially related fields.

To ignore these laws and their implications, invites litigation. Litigation forces retrieval of information. The defendants' case is in peril with an enormous financial exposure if the information can't be found or is not an accurate rendition of the original.

A Computer Archive

Let's consider the characteristics of a good archive.

- There must be significant, extensible storage. Storage that can be expanded and/ or replaced when necessary is essential.
- To address this, a tiered (hierarchical) storage is preferable with cascading performance levels. Further, the ability to add another tier without reloading is a must, as organizations may initially decide to limit their investment. Note that sub-second latency is not usually expected of archives.
- Technologies relating to price/performance of the components vary from direct memory, online disk, alternate sites (online) disk and optical media through to loadable tape / cassettes.
- A strategy must be developed and implemented to cascade the objects to slower, less expensive devices. This activity must be automated; there isn't enough staff to perform any of the actions beyond mounting tapes/cassettes, nor would manual processes be reliable or accurate enough to meet best practice needs and expectations.
- Processes for loading and retrieval suitable to the current and future volumes depending on the request must be automated as well.
- As technology becomes a global requisite, UNCITRAL Model Law, a United Nations set of standards including the modernization and harmonization of rules on international business, will impact archiving. Staying in tune with the standards, now and as they evolve will be a target of businesses.

The Storage Application

Having identified the device technology, there must be concomitant software / firmware to operate the archive to classify, move, index and discover data.

- A robust Workflow that manages the process must be included. Current technology provides the management process which in the Best of Breed includes monitoring screens that graphically show activity with performance related statistics. Jobs underway can be displayed, providing a drill down process to individually view their state.
- International standards will be followed. To this end the International Standards Organization (ISO) has formulated standards. "The initial effort has been the development of a Reference Model for an Open Archival Information System (OAIS). The OAIS Reference Model has been reviewed and pending some editorial updates, it has been approved as an ISO Standard and as a CCSDS Recommendation."
- This covers the delivery to the archive, retrieval and data formats, for only one reason; that someday in future after the current operators, analysts, programmers and managers have retired or simply left, the information can be retrieved with the then current tools. The best situation is an integrated discovery tool with the archive joined to the indexing process.
- During this process, a standardized indexing system is used. The ability to build Metadata, which defines the content in a meaningful way, is crucial, particularly if litigation is involved. When the data is stored, there is no guarantee what information will need to be retrieved nor on what basis.

- Support for extensive format types and data standards is required. Resources such as images, TIFF bit maps, and print streams such as PostScript, PCL and AFPDS may prove difficult to retrieve over time. An archive of PDF documents is advisable. As an example, PDF/A (PDF for archiving), provides an indication of the technical complexities.
  - PDF/A standards require that all resources must be imbedded. No audio or video content, encryption, or executable files is permitted. The Metadata must follow standards. All these restrictions are to increase the likelihood that an archived document can be read successfully decades later (so, no passwords to lose, no exotic video/audio codes required, and no obsolete language interpreters needed). The following details are particularly valuable:
    - PDF/A-1 is a constrained form of Adobe PDF version 1.4 intended to be suitable for long-term preservation of page-oriented documents for which PDF is already being used in practice. The ISO standard [ISO 19005-1:2005] was developed by a working group with representatives from government, industry, and academia and active support from Adobe Systems Incorporated.
  - The PDF/A standards attempt to maximize:
    - device independence,
    - self-containment, and
    - self-documentation.
    - The constraints include:
      - Audio and video content are forbidden,
      - Javascript and executable file launches are prohibited,
      - All fonts must be embedded and also must be legally embeddable for unlimited, universal rendering,
      - Colorspaces specified in a device-independent manner,
      - Encryption is disallowed.
    - Use of standards-based metadata is mandated.
    - The PDF/A-1 standard defines two levels of conformance: conformance level A satisfies all requirements in the specification; level B is a lower level of conformance, "encompassing the requirements of this part of ISO 19005 regarding the visual appearance of electronic documents, but not their structural or semantic properties."
- One must ensure that any new document type can be handled for input and output.
- An organization's requirements will change, but adherence to data standards must remain. If necessary, this may mean reformatting data, recognizing that maintaining the applications that store and retrieve non-standard data may not be indefinitely practicable. Archive and retrieval software selection has decade-plus implications.
- Addressing data compaction is necessary. Technology to reduce repetition or clutter can see reductions reported in excess of 80%. One approach, deduplication, is performed as archiving is taking place. This brings up the danger that deletion at some point in future must not damage referential markers that have been developed for the multiple sets of information.

Lifecycle-based Decision Making

A process known as Records Retention Management is developed by each organization. The objective is to identify the need and value of the information being stored and to develop management rules for each type of document. As a particular set of essential business information, it needs this formalized process.

Lifecycle Embodiment covers object creation, usage, level of access (how important, what retrieval parameters) archival (how long) and finally disposition (deletion). For our purpose, we concentrate on the archival and deletion aspects.

- Determining what to archive requires management attention as the variety of objects expands. Today vast amounts of data are being generated by social media applications such as Facebook, YouTube and Twitter that relate to retail operations where the company's products are assessed in the public domain. This is the operational arena of various forms of analytics, information that has immediacy.
- Using Customer Experience Management techniques the likes and dislikes are evaluated. Does this information have more than a transitory use? Should it end up in a corporate data warehouse, shifted to an archive for lengthy post mortems? This is just one class of information coming in from the public sphere that has chat rooms, formal websites and so on.
- Risk Management techniques are used to assess the information to be stored against criteria including legal and compliance requirements.
- Deletion / data destruction: this brings up the need to define deletion criteria, based on need and legally necessity. In particular, there is an advisory that once the threshold has been reached, the data / objects should be discarded with the exception if litigation requires a hold.

Note that the need to retain information depends on the information and industry. Tax information, depending on the country, may be 6 – 7 years after filing. Insurance records must usually be retained for six years after the end of coverage, which is minimal for an automobile policy, but could be over 100 years for a Life Insurance Policy. In the US, credit card bills must be retained for seven years; in the European Union, it is ten years.

It is essential that the records / information must ultimately be retrievable, which requires and assessment of the life span of data versus the supporting technology.

Disaster Recovery – The Place of Archiving

Disaster Recovery requires an operational restart. Disaster Recovery plans are developed to ensure that the organization has access to all data required to continue business, and that may involve resorting to archives as a last resort. However, an archive doesn't offer the resources that are consistent with Disaster Recovery. It is wise to have a backup of the archive in place.

Best Practices for Archives

What characterizes best practices?

- Governance: Management oversight in establishing the policies that determine the archive parameters.

- Administration: Once policies have been established, classification of the objects must be made and inserted as control over the processes. Polices take security into account, not only protecting the entire archive from improper use or intrusion, but individual's information security such as the medical records enunciated in the USA by HIPAA. Individual's information is complex as jurisdictions vary on requirements. As an example, California, USA considers a person's ZIP (postal) code subject to protection!
- Physical Implementation: An archive is a physical choice; utilities, products or services to deliver the archive function must be chosen. It is strongly recommended that the choice is holistically chosen making sure that the components work together.
- Format: Choice of archiving format; use PDF/A-1 to ensure future data retrievability.

References

Phil Goodwin, How to create a successful data archiving strategy, Storage Digest, January 2014

Lindsay Wyse, Four factors to weigh in planning an analytics big data architecture, Search Business Analytics, July 2013

ISO Archiving Standards, http://nssdc.gsfc.nasa.gov/nost/isoas/ last revised y John Garrett, 2006

Phil Goodwin, Archive It! Storage Magazine Online, June 2013

Gartner, IT Glossary http://www.gartner.com/it-glossary/idars-integrated-document-archive-and-retrieval-system

Technical Information Paper No. 12, Digital-Imaging and Optical Digital Data Disk Storage Systems: Long-Term Access Strategies for Federal Agencies, http://www.archives.gov/preservation/technical/storage-strategy.html, July 1994.

Elings and Waibel, MetaData for All, First Monday Journal, March 2007, http://firstmonday.org/article/view/1628/1543#author

Long term preservation (PDF/A) http://www.digitalpreservation.gov/formats/fdd/fdd000125.shtml

UNCITRAL websites: http://www.uncitral.org/uncitral/en/uncitral_texts/arbitration/1985Model_arbitration.html

Printing Technology

Contributed by Franklin Campbell edp

What is Printing Technology?

To set the context, Printing Technology covers the engineering and execution of getting marks on paper. In the decades since enterprise printing began to evolve the requirements for printing have gone through many changes, including evolutions in how data is developed, generated and integrated into print files. This topic is meant to focus on the physical attributes of printing equipment.

Printing Technology Terminology

There are a variety of technologies in use, each with their own terminology. The following terms and definitions are not comprehensive, but meant to provide the basics to understand the following sections.

Impression/Page/Sheet

In a cut sheet print environment an impression is a single side of a physical piece of paper. Some vendors state their machine speed in impressions per minute. In a continuous web print environment an impression is a printed page image on the web. Remember that print devices capable of multi-up printing can have a higher impression per minute value.

Dots per inch (dpi)

The resolution of the printer. The accurate representation is dots per inch squared because the resolution describes the horizontal and vertical number of dots in a square inch. Resolutions may be the same in horizontal and vertical directions or may be different.

Pixel or pel

A picture element, which is a screen resolution measurement sometimes used to describe print resolution. It is the smallest addressable element of resolution.

Lines per inch

The vertical print speed for impact and other early line-oriented print devices.

Print Media Alignment

The process of ensuring that paper is in alignment with the marking technology. Processes exist for both cut sheet and web fed devices.

Bit depth

The number of bits used to describe a dot.

Anti-aliasing

The process of smoothing perceived jagged edges in type, graphics and images.

Dot size

Measured in Pico liters, this is the amount of ink placed by pulse in inkjet technology.

Dot gain

The percentage that ink spreads away from its original placement.

Points

A point is approximately 1/72 of an inch. It is the common measurement for type sizes.

Simplex/Duplex

Simplex describes printing on a single side of a physical page. Duplex describes printing on both sides of a physical page.

Portrait/Landscape

Portrait describes formatting across the narrow dimension of the page. Landscape describes formatting across the widest dimension of the page.

PPM/IPM

This is a rating of the speed of the device. Pages per minute versus impressions per minute.

Duty Cycle/Monthly Volume

This is the maximum throughput of a print engine over a one month period.

Production Printing Imaging Technology

Electro-Mechanical Technology

Electro-mechanical printers cause a piece of type to impact a ribbon and create marks on paper behind the ribbon. Impact printers are in use today supporting applications that require multi-part form printing and printing in hardened environments where mechanical devices are more appropriate.

The IBM 1403 line printer was introduced as part of the IBM 1401 computer in 1959. The original model could print 600 lines of text per minute and could skip blank lines at up to 75 inches/second. The standard model had 120 print positions, with an additional 12 positions available as an option. A print chain with at least five copies of the character set spun horizontally in front of the ribbon and paper. Hammers struck the paper from behind at exactly the right moment to print a character as it went by. In later models the print chain was replaced by a print train where print slugs were placed in a track instead of being mounted on a chain. The hammer struck the page hard enough that the printer could support applications requiring multi-part carbon form printing.

The standard 1403 chain or train could print 48 different characters: 26 letters, 10 digits, and 12 special characters: &, . - $ * / % # @ ≠ ¤. The ink ribbon was a long roll the width of the print area that was positioned between the print chain and the paper. The roll came in two parts, the feeder roll and take-up roll that was wound and rewound during printing.

The 1403 used fan-folded box fed paper often pre-printed with green bars or gray bars and pre-processed with perforated edges for tractor feeding (pin fed). A carriage control tape or, later, a buffer under program control, specified form length and the form line where printing was to begin so that various paper sizes could be used.

The line mode data format for the 1403 became a standard and still exists today in legacy applications.

As printing volumes increased, various impact technologies were introduced to allow these printers faster speeds. Rather than a fixed print train, the printhead moved across the paper. These technologies include Daisy Wheel (a print disc containing the typeface characters), Dot Matrix (prints each character within a grid), and Line Matrix (prints a page-wide line of dots; it builds up a line of text by printing lines of dots).

Electro-Photographic (EP) Technology

EP print technology is based on an electrically charged drum that is targeted with focused beams of light. Wherever the light hits, the drum loses its electrical charge. When the drum meets a supply of toner, the toner sticks to the non-charged parts of the drum creating the image.

Most toner is a carbon powder and polymer substance. The formulation is unique to the printer, and critical to its performance: key characteristics are granule size, melting point and polymer type. For most toner-based printing the fuser heats the toner and roller pressure binds the toner to the page. In Electron Beam Imaging the process is slightly different; the image is transferred to the drum using electron particles instead of heat.

EP Pioneers

Early EP devices from Xerox include the Xerox 1200 Computer Printing System. Introduced in 1973 this solution combined both copying technology and a computer. It printed at 4,000 lines per minute or about 60ppm. The follow-on product was the 9700 production laser printing system in 1977. This cut sheet laser-based printing system operated at 120 impressions per minute.

In 1975 IBM introduced the 3800 continuous forms printer allowing it to handle the same paper media as the 1403, but at 180 impressions per minute.

EP Process

Electrophotographic process in laser printers, involves six basic steps:

1. A photosensitive surface (photoconductor) is uniformly charged with static electricity by a corona discharge.
2. The charged photoconductor is exposed to an optical image through light to discharge it selectively forming a latent or invisible image. This is called exposition.
3. Development of the latent image is achieved by spreading toner, a fine powder, over the surface. This adheres only to the charged areas, making the latent image visible.

4. An electrostatic field transfers the developed image from the photosensitive surface to a sheet of paper.
5. The transferred image is fixed permanently to the paper, by fusing the toner with pressure and heat.
6. The excess toner and electrostatic charges are cleaned from the photoconductor to make it ready for the next cycle.

Laser printers and LED (laser-emitting diode) printers typically offer the highest resolution. The difference between them is the method of exposition or formation of the latent image

EP Imaging Technology

EP Printers are characterized by:

- **the imaging technology used:** laser, LED, or electron beam imaging,
- **the paper transport type:** cut sheet or continuous forms, and
- **color capability:** mono, highlight, full color.

Laser Printers[1]

Laser printers use electrical charges to write images on a drum where toner will adhere until transferred to the intended substrate, usually paper. You may encounter the terms Corona Wire or Charged Roller. These are the mechanisms for putting a positive charge onto the drum to prepare it for laser imagining. The laser causes the points on the drum where the beam hits to discharge preparing the drum to accept toner.

With the toner now on the drum it is transferred to the substrate and then the drum continues to revolve around to the charging area. Transfer may be direct or to a transfer belt and then to the substrate.

LED Printers[2]

LED printheads use an array of Light Emitting Diodes (LEDs) to flash the image on the drum. Depending on the timing and intensity of the LED the image quality can vary, though modern LED printheads have mastered the management of these mechanisms. Just as in the Laser printer, the drum begin it's revolution positively charged and the LEDs image the drum to prepare it for the toner. The toner image is tranferred to the substrate, usually paper. Transfer may be direct or to a transfer belt and then to the substrate.

Electron Beam Imaging

Delphax Systems was founded in 1980 in Canada. Delphax developed a cold fusing technology using electrons to image the drum. Patented as Electron Beam Imaging (EBI) it formed the core of technology for printing at high speed on a variety of substrates. EBI technology requires 30% fewer components than laser printers.

1 Based on information found at http://mimech.com/printers/laser-printer-technology.asp and http://computer.howstuffworks.com/laser-printer2.htm
2 Based on informaiton found at mimech.com/printers/laser-printer-technology.asp and http://www.office.xerox.com/latest/W74WP-01U.pdf

In 1997, Xerox owned one-third of Delphax shares and purchased the remaining two-thirds to gain control of the company. In 2001, Xerox sold its interests in Delphax to Check Technologies. The combined company is now called Delphax Technologies.

EBI advantages include:

- reliability,
- speed,
- uniform and durable imaging,
- enhanced dot uniformity, and
- better image control.

The following steps take place in the EBI process:

1. Image placement.
2. Toner application.
3. Fusing:
   - pressure fusing,
   - radiant heat fusing.
4. Toner and charge removal.

Inkjet Technology

Inkjet printing involves the use of jetting modules to release ink that lands on the substrate. Inkjet technologies are used in individual print modules that may be mounted on analog printing presses, inserting equipment or other finishing equipment to add variable data or they may be incorporated into paper transport devices to create inkjet printing systems.

There are many engineering options for getting the ink to the paper. The two main technologies are drop-on-demand (DOD) technology, where each drop of ink is released on command for specific placement to form characters and images, or continuous inkjet technology (CIJ), where ink rains from the print module but only designated droplets land on the page. Remaining droplets are pulled back to recirculation systems for reuse.

Continuous Inkjet

Continuous Inkjet technology evolved from work done in the 1860s by Lord Kelvin. While Siemens introduced the first commercial inkjet printing systems in the 1950s for medical recording, today Kodak is the primary vendor in the CIJ inkjet space. Beginning with technology acquired from Mead Paper, Kodak marketed inkjet modules and printers under the Diconix brand. That division was sold to Scitex to form Scitex Digital Printing, which was sold back to Kodak in 2005. The core technology remained CIJ-based print modules and web presses.

The first generation of their CIJ used electrical charges to cause ink to drop to the paper or divert to the recycling system. The Prosper press series uses air deflection techniques to manage droplets. In both cases Kodak is the designer and manufacturer of their print modules.

Inks may be water-based or solvent-based and may be dye-based, pigment-based or a blend depending on the target application. For most enterprise printing ink is generally water-based dye or pigment.

Drop-on-Demand Inkjet

Most of the other vendors in the inkjet market sell Drop-on-Demand systems. Canon, HP, Kodak, Ricoh and Xerox have products in this market, along with a variety of companies selling proprietary solutions such as the RR Donnelly ProteusJet.

DOD printheads use Thermal jetting (TIJ), where ink is heated causing it to expand and jet through a nozzle, or Piezoelectric jetting (PIJ), where a charge causes the ink to jet through a nozzle.

While HP designs and manufacturers their own TIJ printheads most other vendors use PIJ printheads sourced from the Japanese technology companies Seiko-Epson, Kyocera and Panasonic. Vendors using printheads purchased from the Japanese Original Equipment Manufacturers (OEM) are using PIJ technology. Most consumer inkjet printers, including those from Canon, Hewlett-Packard, and Lexmark, use the thermal inkjet process.

Print Controller Architecture

The Print Controller is designed to ensure that each component of the printing system performs its function at the right point in the print job.

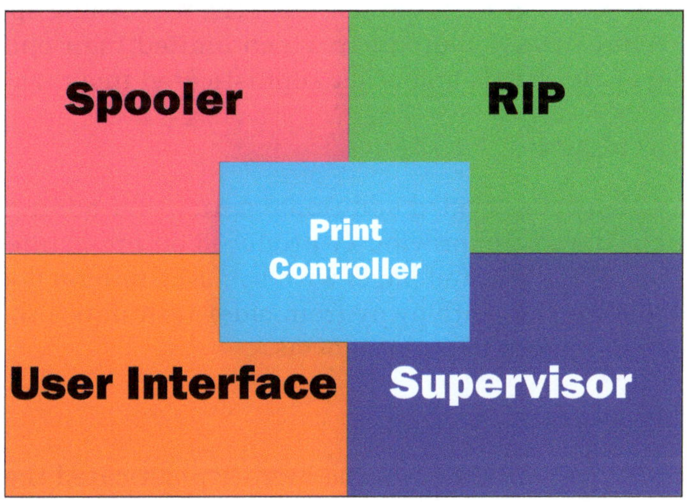

It has several elements:

Print Spooler:

- responsible for receiving jobs from Network via various protocols, and
- attaches to print server.

Print RIP:

- responsible for imaging pages,

- converts print streams into raster images, and
- machine image controller (MIC).

Print Supervisor:

- responsible for controlling print run and responding to operator commands,
- manages servo-motors in paper path,
- monitors paper jam sensor,
- monitors photoconductor charge strength,
- error control, and
- reporting.

User Interface:

Responsible for communicating to the Printing System:

- micro code or programming,
- provides programming for control unit, and
- manages printer and its functions.

RIP (Raster Image Processing) Technology

The job of the RIP is to digest the print stream and build the raster or bitmap representation of a page at a speed that is as fast as or faster than the speed of the print engine. First generation RIPs did not have the processing speed or memory capabilities of current generation products which limited their options for print directions, printing on diagonals and other sophisticated imaging.

Electronic Overlay Memory

Electronic Overlay Memory provided additional options for adding electronic images of content that might have been previously produced on pre-printed forms by creating a single matrix the size of a printed page, with some 11 million dots. The Overlay Memory provided a migration path away from older techniques that involved using film to flash images onto the paper to replace forms.

RIP Subsystem Specialization

As the capabilities of print streams became more sophisticated the functions of the RIP had to follow along. As Document Objects changed such as bitmap fonts to outline fonts, RIPs added sub-processing functions to handle building the bit map representation of the font on the fly. When vector graphics were added, the RIP added a module to process 2D vector graphics. To handle color processing of four different bitmaps to represent CYMK, RIPs added another subsystem. The end goal was to produce a page at least as fast as the print engine. When a RIP cannot keep up with the print engine undesirable results occur. Either the print engine stops, and waits for more pages to be buffered, or in the worst case the engine starts and stops, sometimes called clutching. It is not good for the printer engine and is usually a sign something is not going as expected inside the RIP.

Cut Sheet Printing Systems

Cut sheet printers enabled a wider variety of mixed media types within a job. These machines are used in mid-to-high volume shops enabling applications that required multiple colored papers, sizes, tabs, and NCR (carbonless multi-part) paper media. Their speed ranged initially up to 120 pages per minute, and today cut sheet solutions exceed 300 ppm. Duplexing was accomplished by using a turnover mechanism, and running each page through a second pass. Current technologies provide simultaneous duplexing using two imaging units to image the page in a single path. Typically, the shorter the paper path, the more reliable and less jamming occurs. Since paper flies through the engine, paper acclimatization and environmental conditioning of a cut sheet print room are critical.

Cut Sheet Technology Components

Component Name	Area	Description
Input Drawer	Cut sheet paper input	Manual and/or automated power drawers Multiple drawers Multiple paper sizes or media types Also called a hopper
Picker/ Paper Transport Unit	Imput	Unit that pulls paper into paper path
Marking Engine	Print Engine	Builds and transfers image to media
Bypass/Sample Tray	Print Engine	Tray used to make samples from the print run for quality checks.
Stacker	Output	Straight or off-set stacking of output pages
Interposer	Post Print Engine	Used to pick pre-printed material possibly color to add into the document
Peripheral Devices Folder Job Finisher Saddle-Stitch/ Booklet Maker Perfect Binder Professional Punch/3HP	Post Print Engine	Various Folds Single/Double Stitch Offset Stack Booklets Perfect Bound Simple or GBC Punching

Continuous Forms Printing Systems

Continuous Forms printers differ from cut sheet print devices in that the paper is pulled through the printer from a box or roll of paper. This limits printing to a single media type without changing out the roll. Due to the method of pulling paper through the printer, these devices can print at much higher speeds and typically experience few paper jams than cut sheet devices.

One limitation of this design is that to print duplex, a second engine is needed. The first engine images the front side, and the second engine images the back. There are tri-plex configurations where a second color can be added.

As these machines became faster, more sophisticated pre/post-processing equipment was created. Unwinders capable of holding 50 inch diameter rolls of paper were used instead of boxed stationery. A variety of post processing equipment was developed to meet various needs. Rewinders allowed the paper to be rewound onto a roll after printing; these rolls could be taken and fed into inserting and other finishing equipment. Roll-to-cut or Cutter/Trimmer/Stacker was another popular variant, as well as roll-to-fold.

While the original continuous forms machine required pin/tractor fed paper, newer models are offered in pinless options. The main advantage is the ability to expand the print width and not having to strip off the ½ inch on each side during the inserting process, producing less waste and eliminating a device on the inserter to dispose of the trim waste.

Component Name	Area	Description
Un-winder	Input – Variances: Core Size, Roll Width	Handles feeding paper roll into the printer; 1 to 3 hours Print Time between roll changes
Auto Loader	Input	Automatic threading of paper through print engines
Marking Engine	Middle	Area where the page image is built and transferred to the media
Pressure Roller	Input	Pulls paper through without holes
Tractor Feed	Throughout	Pulls holes along paper sides
Pinless	Throughout	Eliminate disposal of trim waste
Cutter / Trimmer / Stacker	Output (CTS)	Cuts rolls into sheets of paper (usually A4 or Letter) Sorts into sequence

Component Name	Area	Description
Folder / Separator	Output	Creases and folds papers Separates paper at print job boundaries
Rewinder	Output	Used to create Roll-to-Roll processing solutions

Resolution vs. Quality

One approach to getting higher image quality when using legacy 300 dpi bitmap fonts is to perform anti-aliasing or edge enhancement. This works best with characters with curved or diagonal edges and helps to smooth the appearance of characters on the page.

MICR[3]

Magnetic ink character recognition (MICR) is a character-recognition technology used in direct mail marketing and financial services, as well in check imaging to ease the processing and clearance of checks and other documents. The most common application is the MICR encoding is called the MICR line and is found at the bottom of checks and other vouchers. The technology allows MICR readers to scan and read the information directly into a data-collection device. Unlike barcodes and similar technologies, MICR characters can be read easily by humans.

The two most common MICR fonts are the 14-character E-13B and the 15-character CMC-7. The E-13B font is part of the ISO 1004:199 but CMC-7 is most commonly found in Europe. MICR solutions use MICR ink that contains pieces of metal which become magnetized when printed.

The MICR E-13B font is the standard in Australia, Canada, the United Kingdom, the United States, and other countries. The 13 in the font name comes from the 0.013-inch grid used to design it. In addition to numeric characters it includes the ⑆ (transit: used to delimit a bank branch routing transit number), ⑇ (amount: used to delimit a transaction amount), ⑈ (on-us: used to delimit a customer account number), and ⑉ (dash: used to delimit parts of numbers—e.g., routing numbers or account numbers). Major European countries, including France and Italy, use the CMC-7 font, developed by Groupe Bull in 1957. CMC 7 MICR lines are usually printed as close to 8 characters per inch as possible to accommodate the MICR readers. The font is available in a variety of resolutions.

MICR Quality

The quality of MICR printing must adhere to national and international standards established for MICR documents. In the United States, these standards are developed by the Accredited Standards Committee (ASC) on Financial Services, X9, operating under the procedures of the American National Standards Institute (ANSI). The key elements of performance for MICR toner include the following:

- **Character Signal Strength:** A measurement of the magnetic property of the toner.

3 http://www.sourcetech.com/micr-toner-supplies/quality-micr-toner

- **Adhesion:** The ability to adhere or fuse the toner to the paper stock.
- **Durability:** The performance of the toner through multiple passes in reader/ sorters.
- **Contamination:** The effect of MICR toner on the base printer hardware.

Color Printing

Color Printing for Transaction Applications is either spot color or full color. Spot Color provides black plus one other color option. Full Color Technology may be based on red, green and blue plus black (RGB), or cyan, magenta and yellow plus black (CYMK).

EP Color

Due to the physical limitations of EP imaging engines it is necessary to make multiple passes over the page depositing the colors that build up to render the author/designer intent.

Inkjet Color

In inkjet technology a full color impression can be achieved in a single pass through the printer. Arrays of printheads deliver the various colored inks effectively in a simultaneous process.

Density

Color density is much like greyscale and impacts the gamut of colors that can be represented on a page. A very low color density like 4 bits may not accurately represent the colors of the original full color image. However, there are eventually diminishing returns with higher and higher density levels.

All printers have a limit to the color spectrum or envelope of colors that they are able to produce. At some point higher color densities may exceed the envelope of the print engine. There is also a cost in the volume of data to represent these higher densities, and a consequential impact on the RIP processing time.

Color Density Impact:

- 8-bit results in 256 color combinations,
- 24-bit results in 16.7 million combinations, and
- 32-bit results in 16.7 million colors plus indicators for opacity.

Ink

Inks fall into two broad categories: Pigment-based inks that stay on the paper surface and dye inks that are absorbed into the paper. Each manufacturer has their own process for formulating ink that is appropriate to the jetting technology and printing intent.

Water-based dye inks and pigment inks are the most common in the enterprise printing environment, while solvent-based pigment inks are commonly found in industrial inkjet environments. The key characteristics of inks used in the enterprise print space are that they must adhere to the paper to survive finishing and mailing operations and they must provide sufficient image quality to be acceptable to the brand owners. The paper used is a key component of how well the image quality is perceived since how the ink is absorbed by and adhered to the paper surface is a major determining factor in perceived print quality.

Over the last half century the ability to create inkjet droplets that are smaller and to use ink that is more refined has led to the ability to create inkjet print devices capable of higher resolutions. Controlling the size of the droplets, and therefore the amount of ink dropped onto the paper, impacts drying times and the precision of text edges and grain in images. The paper selected for printing also impacts the quality of print since the interaction of the paper and ink, and the rate at which the ink is absorbed into the paper (porosity) impacts the perceived clarity of the print.

Another facet that impacts print quality in inkjet is the concept of dot gain. This is the amount of spreading from the initial droplet that occurs when the ink hits the paper. This is a common discussion in offset printing, but only entered the digital printing discussion with the rise of inkjet since toner-based printing does not experience dot gain.

Depending on the size of the droplets, measured in pico liters, and the porosity of the paper, dot gain may be limited or the ink may expand considerably beyond the droplet boundary. In all inkjet applications the combination of ink and paper determines the dot gain, and therefore the perceived clarity of the print, so testing for acceptability by the end customer is essential.

Inserting

Contributed by Rebecca Rodgers m- edp

What is Inserting?

The industry is continually looking to reduce machine stoppages, reduce waste and expenses, maximize operator productivity, increase integrity of the finished mail piece and meet ever increasing service level demands. And, inserting has always been the Achilles heel of the transaction document production line.

Most seasoned document production people know that the majority of the document production problems happen at the time of inserting. This step is the most mechanical; often the least automated, and subject to operator error or malice. Every document that jams on the inserter has to be reprinted. Every page that 'sticks' to someone else's document or ends up in an unintended envelope could cause extensive damage resulting in litigation that tarnishes the organization's reputation / image.

A typical mechanical inserting system may soil (damage) 1 in 10-50000 mailpieces. If you produce 1,000,000 mail-pieces you may lose 100 at this step.

The areas discussed in this section include what should be considered for successful production inserting in the transaction document space, the sub-components of a production inserter, use of barcodes to manage the process, software-based / file-based document control, and touches on envelope (finished mail-piece) ink-jet printing.

In the Transaction Document world, inserting refers to the manual or mechanical production process step that collates, folds and stuffs documents into envelopes.

Inserting equipment can be described as mechanical machines driven by software programs to automate and integrate the collation, folding and stuffing of documents into envelopes to create the mail piece. This equipment uses a defined mechanical path and operator set-up designed to reduce the time and effort it takes to stuff large volume mail. In other words, the equipment ingests printed paper into a conveyor system either from stacked single sheets or rolls, folds documents and additional material into a specific size and inserts it into an appropriate envelope.

Handling complex transaction applications with variable page counts, selective (dynamic) inserting and matching of the transaction mail piece contents must be secure and requires the highest levels of integrity. It cannot be compared with the handling of direct mail.

Today's high volume inserting environments have sophisticated software controlling each step of the inserting process and tracking the document to mail piece construction using (typically) 3 of 9 and / or 2D bar codes. This is what is meant by software-based / file-based document control and requires the use of insert control files (ICF). Some organizations have such software to enable the fundamentals for the ADF (Automated Document Factory), while other organizations have integrated the software into their workflow management environment and tools to enable file-based transaction document production management. The inserting production process is further explained in the section that describes the insert stations.

Terms Used in Inserting

- **Mail inserting:** action of inserting mail into envelopes.
- **Letter Shop:** facility handling mail inserting / stuffing and preparation.
- **Stuffing / mail stuffing:** action of inserting mail into envelopes.
- **Intelligent inserting:** action of inserting mail into envelopes using sophisticated mechanical equipment.
- **Marketing inserts:** supplemental information accompanying the original document in the envelope.
- **Onserts:** supplemental information printed on the primary document), could be a:
    - **Marketing message:** a promotional message, a reminder, an offer or other such message on the primary document which can be targeted to a specific person or group of people a.k.a. demographic or geographic region for example. The message can be created by the client and driven by data / flags in the data file.
    - **Regulatory message:** required by a regulating authority.
- **Flat inserting:** action of inserting mail into large 9 x 12 envelopes without folding.
- **Business reply envelope (BRE):** the smaller envelope accompanying the document, intended for the recipient to return mail to the sender.
- **Divert:** action of diverting particular mail pieces from the bulk of the mail for specific purposes.
- **Postage indicia:** the frank / postage mark showing the postal service customer account number to be charged that is either pre-printed on the envelope or imprinted during the inserting process.
- **Dynamic inserts:** supplemental information accompanying the original document in the envelope that can be different in each mail piece depending on the business rules driving the insert stations.
- **Selective inserts:** supplemental information accompanying the original document in the envelope that is selected (turned on or off) for each mail piece depending on the business rules driving the insert stations.
- **Machine inserting:** action of inserting via machine versus manually.
- **Hand stuffing / hand inserting / manual inserting:** action of inserting / stuffing by hand.
- **Insert Control File or ICF:** a file used in file based processing environments containing system controls, that is electronically sent to inserter controller to instruct the inserter on how to process the physical pages and create the mail pieces (sequence, how many pages per mail piece, which inserts to use etc.).
- **OMR:** an older rudimentary mark on the document consisting of a few horizontal lines down the margins of the page to tell the inserter which is the first page, last page or whether to include an insert. Designed for line printing but still in use today.

- Barcodes, such as 3 of 9 and 2D (a 3 of 9 barcode or 'picket fence' barcode looks like a UPC on retail products and contains a sequence of bars which describe to the inserter, the document, sequence, number of pages, inserts required etc. and can be quite large and intrusive. A 2D barcode is a box-like bar code, higher density than 3 of 9 and can contain much more information about the document, its content and lifecycle. Both support document reconciliation and audit)

Inserting and the Automated Document Factory (ADF) Concept

Inserting is one of the last manufacturing steps before physical distribution. Ensuring that the right mail piece and content gets to the right person at the required time involves all of the prior steps in the document creation process. The requirement is to build and produce the best possible product and track what will be required at the insertion stage so that the final product is error free, risk is mitigated and ultimately result in accurate inserting.

It should be noted that the same considerations must apply all the way from the business requirements through to the production of the electronic / physical document. Errors in physical and electronic distribution are equally damaging.

Document integrity considerations are required in document creation for both physical and electronic production; from variable and fixed data coming together accurately and completely, through building the physical /printed document (composition and emitting a clean print stream). When emitting the electronic document stream for an ePresentation service there is also the requirement for accurate tracking and reconciliation processes along with sophisticated web based security to ensure that the intended recipient is the only recipient who has access to their transaction documents on line.

The creation, composition and output have been described in other sections. Successful inserting is the culmination of upstream production considerations; in summary and at a high level, inserting must be considered and planned for in all the following stages:

- Requirements Gathering
- Business Analysis
- Technical Analysis
- Stakeholder Adoption
- Design
- Architecture
- Development
- Testing and Quality
- Production
- Maintenance
- Decommission

Besides all the production phase requirements described previously, the requirements specific to inserting that each of the above phases need to consider include the following points.

Requirements Gathering

Requirements gathering must identify all the needs of the client, the job, the document lifecycle, the support functions (e.g. customer service / call center), the recipient , the channel and the call to action lifecycle:

- what the output will look like, what size media / stock will it use,
- if the output is to be folded or inserted flat (e.g. could be size related or may only be a few pages, but could be a high value wealth management statement that the business wants to present without folding),
- the expected number of pages per envelope, as this affects the fold and what can fit in an envelope (e.g. 1-6 pages in a #10, 7-9 pages in a 6x9, 10-30 pages in a flat (9x12)),
- what envelope will be used (standard / custom),
- what inserts / onserts / attachments or other materials will accompany (or be part of) the primary inserted document and when, including business reply envelope (BRE),
- if the inserts be dynamic (in that there could be a different variety for each recipient, based on data triggers),
- what business rules will drive the inserts and which business rules will be driven by the source data or other reference files (e.g. geographic or demographic flags that dictate the inserts required),
- what file split requirements exist (convenience (volume) splits, regional, postage etc.),
- any additional sorting required by the Operator,
- the scope of distribution (domestic, international, exclusions. Non-domestic mail will need to be diverted at the inserter if it is printed in the same file as domestic mail, as there is a different cost and handling requirement),
- what methods / channels will be used (postal service, courier, internal/branch mail, special handling),
- the expected cycles and volumes, frequency and exceptions,
- the expected production peak / low times and the impact on volumes and inserts,
- the requirement for handling return mail,
- storage requirements associated with pre-production (e.g. inserts/additional stocks) or post production (e.g. hold before shipping),
- what distribution labeling is required,
- any secondary instructions be sent with the job instructions,
- what test cycles are expected,

- if production materials will be available for testingm
- the expected service levelsm and
- what reporting is required from the system or the Operator procedures.

It is important to note that the slightest change to the current process / environment should kick off an impact assessment to the entire production process. Knowing as much up front as possible allows the development and composition processes to be more efficient.

Business Analysis

Business analysis must identify all the elements that incur costs / expenses in order to accurately predict what the job can be priced at:

- what training will be required,
- how long will it take to run the job/s,
- if it take more operators to run this job than exist,
- what exceptions might incur additional cost,
- storage space,
- materials usage and inventory management impacts,
- effort required to complete all tasks including validation, reconciliation, reporting and manual sorting, and
- additional parts, modules, equipment,
- if all the job elements fit within standard practice / process, if not what is custom and how much extra will it cost (e.g. is inserter programming, professional services support etc. required).

Technical Analysis

Technical analysis must focus on the technical solution and identify all the components of the environment and the impacts to current state. It must ensure that the current technology is assessed for capacity and capability, required enhancements or replacement technology, the impact to processes and practices, and the impact to the resource pool. This analysis is not providing costs for the solution, but identifying the elements that need to be assessed for cost by the Business Analysis process, including:

- What equipment will be used to produce the job, is there capacity at the expected times.
- If this equipment produces the required volumes (including peak) in the service level window.
- Iif the current inserter solution has the programming capabilities required for the job; if not, what needs to change and the impact.
- Any additional factors required to produce this job (e.g. additional folding plate, more insert stations, additional cameras/readers, different input station, changes to divert station.

- If the reporting requirements be met with the information / systems that will be gathered on the job.
- How re-makes / re-prints will be produced and by whom.
- What bar code will be used and if it meets the production specification.
- The location of the bar code and if it meets the production specification.
- Insert control file content and if it meets the production specification.
- Job instruction content and if it meets the production and operator specifications.
- Iif the system/s produce the required reporting.

Stakeholder Adoption

Stakeholder adoption must confirm understanding and acceptance of the overall solution and the components within the solution. This includes the solution costs, the solution impacts and the solution plan, as well as a review and approval process for items such as:

- changes to current equipment,
- changes to current workflow,
- changes to current procedures,
- changes to reporting,
- changes to communications,
- changes / impacts to other services, and
- changes / impacts to other service levels.

Design

Design must consider all of the elements of presenting the document to the recipient and all of the support functions within the document lifecycle and other document lifecycles prompted by the document call to action, including:

- the use of white space,
- branding and standards,
- all channels of delivery and receipt,
- what business rules will drive exceptions, such as programming the bar code or other insert identifier must consider if the marketing message is not present due to upstream processing restrictions (e.g. the message won't fit in the white space available) should the insert still apply, or if the language flag is missing what is the default language for inserts when multiple languages are being processed,
- placement of integrity marks such as bar codes, so as not to impede the presentation of the document content or address block AND place it in a standard location to reduce set up time on the inserter, providing enough white space around the code so it can be read accurately and consistently, and

- envelope window alignment - document content must not show in the window, only the address and required / approved integrity marks.

Architecture

Architecture, as it pertains to inserting must consider how all the technical components 'hang' together without negatively impacting current state and how the new solution will best fit the environment:

- What inserting connectivity requirements are there, are any enhancements to the network or servers required to 'talk' to the inserter.
- If new inserting PC's / servers are required, how will they be implemented.
- What systems will be receiving data and how will that data get the required inserting programs and processes.
- Does the current environment support connecting the solution in a standard configuration, if not, how will it be accommodated (e.g. is the job instruction / control file coming from the current solution or is being supplied by the client or other system, how will this information make it into a standard ICF.
- How the systems and the jobs run in a business continuity / disaster scenario, and what is required at a secondary site.

Development

Development must receive clear and concise specifications from the requirements phase, business and technical analysis and architecture phase. Development specific to inserting includes such items as:

- control files to run the inserter programs,
- providing information that will support the creation of an electronic job instruction for equipment Operators or manual insertionm, and
- ensuring that the bar code or other integrity marks (e.g. human readable instruction (HRI) are standard and in a standard location and can easily be read by the inserter, or by people in the case of manual insertion.

Testing and Quality

Testing and Quality must include all the individual components being tested thoroughly plus end to end testing to ensure that all the components 'hang' together. At a minimum, inserter testing (prior to production) should include:

- actual production equipment tests, including 'stress' test (a.k.a. high volume test),
- actual production stocks (document, inserts, envelopes - both outer envelope and reply envelope / BRE),
- bar code and other integrity mark readability tests (camera, scanner, reader),
- bar code success in consolidating accurately the document and relevant inserts,
- quality of the fold,

- quality of the document as it reaches the envelope (ensure the edge are intact after any turns / accumulation,
- page imposition and order to ensure it feeds in the right order to accumulate and match the control file and required integrity,
- ability to read the address in the window in all scenarios (e.g. single sheet to maximum number of sheets and inserts in an envelope),
- ability to match the mail piece content to the address if imprinting the address on a closed envelope,
- time and effort on the inserter,
- file splits are accurate (e.g. volume / convenience splits, regional / geographic splits etc.),
- divert mail pieces (diverts) are accurately and appropriately diverted ability to report (for internal and external purposes) as required,
- exception handling,
- operator knowledge and comfort, and
- business continuity.

Production

Production steps must include checking continuously the quality of the run, either by closely monitoring the system checks and error notifications or by regular manual checking of mail pieces if running a non- file based environment. In a manual workflow the specific number depends on the organizational practices and job volume. For example, prior to running the job the operator will generally check that:

- the job, the envelope/s and inserts required are accurate and available,
- the job has been delivered and appears intact (e.g. does not appear to have been dropped or shifted or re-ordered),
- the start and end sequence numbers are present and accounted for (if the job is very large the end sequence number may be on a different cart or still at the printer, this end sequence refers to the last sequence number in the batch and on the job instruction / docket),
- the inserter program is pending / ready to run,
- the quality of the print job (quick visual check of visible (on top) pages appears good, and
- the equipment has been cleaned, set up according to job instructions and is ready to go.

While running the job the Operator will generally:

- monitor the input station to ensure that it has a constant feed to maintain productivity,
- monitor the insert stations to ensure they are re-filled with the correct inserts,
- monitor the screen, watching for any errors / unusual diverts,

- listen for alarms indicating intervention or a check required,
- monitor the water reservoir to ensure the moistener functions and envelopes are sealed, and
- perform visual checks of the line and physical checks of the finished mail piece to ensure consistent quality.

Tracking, Validation, Reconciliation processes must capture:

- that page numbers and sequence numbers are all produced accurately,
- that complete 'sets' of documents / mail pieces are produced and accounted for,
- time and date for production steps (usually captured by the inserter system), and
- validation of all content and volumes.

Maintenance

Maintenance of the inserters must include a Preventative Maintenance (PM) schedule. Preventative maintenance cycles (PM's) are important to manage as inserting equipment needs regular cleaning and / or parts replacement. PM's are usually done by the vendor and performed in a pre-determined 'quiet' period as it usually entails the equipment being stripped down more than during general maintenance.

As the equipment runs at high speeds using friction, air, suction feeds, conveyors and other process, there is a lot of physical wear and tear. The equipment is finely tuned to feed and accurately separate paper (handling single and multiple sheets at the various stages) and read the integrity marks using scanners / cameras, meaning that keeping it clean and clear of paper dust and other debris is crucial to a smooth running shift. Well-trained Operators can clean and maintain the basics, but the Vendor service technicians should also be on a regular general maintenance schedule and a regular parts replacement schedule, which should all be included in the Vendor Service Agreement and reviewed regularly at Vendor Performance Reviews / meetings.

General maintenance is performed constantly and will include at a minimum:

- operator cleaning during or after each shift, and
- operator replacement of easily replaceable parts that wear out between service calls or PM's

Decommission

Decommissioning the job or the production equipment must include impact assessment, for example:

- how is business continuity impacted,
- how are the Operators affected,
- what peripheral equipment is impacted,
- what support functions are affected,
- are there any equipment disposal considerations (e.g. environmental, financial, logistical etc.),

- if this job doesn't run what system impacts are there,
- what is the impact to the supply chain (e.g. stock and related), and
- financial impact.

Inserter Equipment Components and Sub-components

The equipment components or modular devices that are integrated to create an inserting system are called stations. Following is an example of an inserter configuration. Depending on operating space, number of operators, input station requirements and equipment manufacturer, the equipment layout is configured to meet the requirements to optimize the environment.

This diagram is not intended to represent the best configuration or show preference to a vendor or model, but is meant to generally illustrate the typical components of a production inserter.

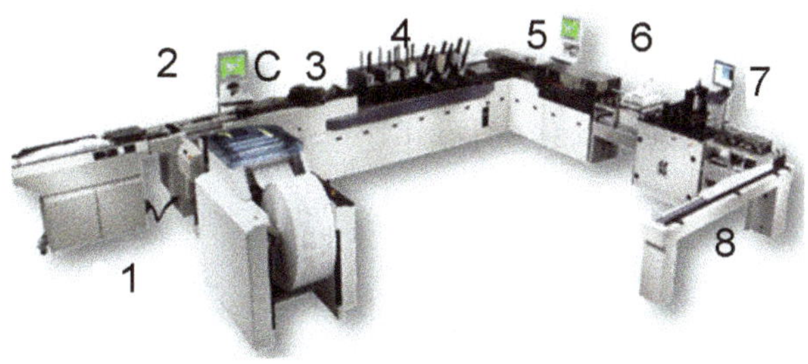

C - The Controller is the inserting 'brain' and contains the necessary intelligence to operate all the inserter components

The controller supports individual job programming which:

- allows the operator to easily repeat set ups,
- provides customized programming to meet specific application requirements, typically used for repeat and complex jobs,
- controls the job run,
- provides audit capability,
- provides error reports,
- informs the operator where and what the error is,
- tracks system performance,
- keeps track of the individual pages / inserts and the finished mail piece,
- can be available locally at the inserter or at a remote terminal,
- provides postal manifest input (ESOM) / mail-piece calculations, and
- can interface with a charge back tool.

1 - The Input Station / conveyor paper handling module which feeds documents to inserter.

If the printing process has produced cut sheet output (single sheets with simplex or duplex impressions) the sheets are laid out in a shingled manner on the feed conveyor belt at the input station, requiring sheet-feeder technology.

If the printing process has produced the output on a roll then the insert equipment requires additional cutting and separating modules to ingest the job and prepare it for the accumulator step.

The pages need to be in the same order as the ICF (jobs are usually pre-programmed at the inserter to recognize the sequence of each document set and the entire job) and / or the job instructions for the Operator will identify the order in which they were printed recognizing the input and collating requirements of the inserting process.

- continuous feed – web forms conveyor:
  - single forms web (roll),
  - merge two webs (roll).
- the job is fed head first or foot first depending of the design and composition:
  - type of application,
  - type of envelope (window or no window),
  - type of fold,
  - location of integrity marks.
- bursts and separates input if perforated and / or needs the tractor (roll paper feed) holes removed,
- cuts and separates input if not perforated and / or needs the tractor holes removed:
  - cuts the roll to desired sheet size,
  - separates ready for the accumulator step.

Obviously, roll input to an inserter offers higher productivity due to less frequent stopping / loading and fanning of the cut sheet input, but the business needs and the printer used will determine the required equipment configuration.

2 - Accumulator Unit

Both roll and cut sheet input needs to be accumulated to create the document set, unless the application is a single page. The accumulator unit:

- assembles sheets into the document set:
  - using integrity marks,
  - ensures accuracy of collation,
  - scans each sheet,
  - bar code, OMR information is captured for validation, reconciliation and audit.

3 - Folder

The folding unit will typically have 2 folding plates to meet standard folding requirements. Usually the organization has a set of inserting standards in place to ensure that the address block is consistent and the folding set up is limited. More plates may be required for more complex folds (not typical in transaction document production).

This step:

- folds the document set (which can be a single page),
- multiple pages are married together:
  - based on accumulated sheets using integrity marks,
  - scanners / cameras check the accuracy of collation using the integrity marks.
- standard sheet folders in a typical transaction document production environment fold 8.5 x 11" or 8.5 x 14".

There are several common folds for transaction documents driven by business requirements for the envelope style and size and sometimes the document purpose. Listed below are common horizontal or vertical single or double folds:

- C fold[1]

- fanfold / N fold / Z fold / accordion fold

- half / V fold

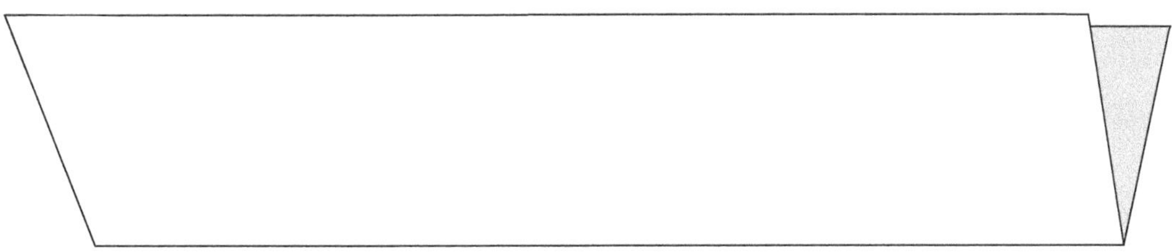

1 Fold images courtesy of FoldFactory.com and used with permission.

- end fold

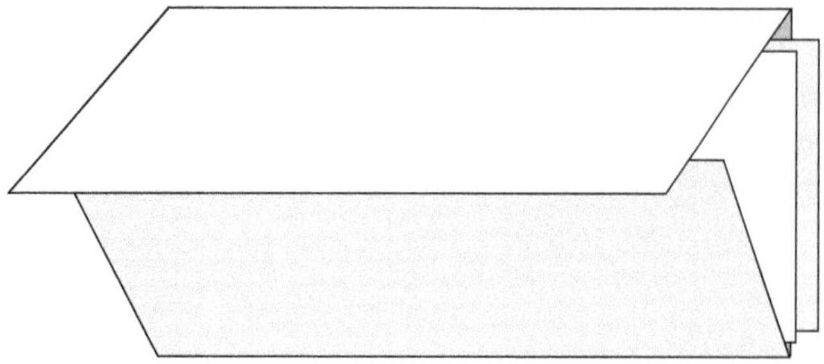

4 - Insert Module / Station

This step in the process creates the finished mail piece content by marrying the document, inserts (including a BRE if used) and the outer envelope to create the mail piece by folding all the relevant sheets together adding the required additional printed materials (inserts, BRE) and stuffing all the consolidated materials into the outer envelope for mailing.

More commonly now the inserts themselves have integrity marks to ensure that the right insert is used with the right job and to add to the validation, reconciliation and audit process.

Often the individual 'feeder units' are also referred to as insert stations. This station consolidates the document set that has been accumulated thus far and the inserts from any of the individual feeders, and places the entire set into the outer (mail) envelope:

- the envelope arrives here typically with the front side down & flap open (this can differ),
- the envelope opener:
  - opens the envelope (blast of air opens the throat),
  - slides the contents inside the envelope using a guide, and
  - pushes the envelope into the next station called the Moistener, Closer, Sealer (MCSE) for sealing.
- there can be multiple feeder units, 4- 6 is the average (but can be up to 8 for large mailers), and
- the inserts can be marketing, legal / program compliance inserts and usually include a business reply/return envelopes (BRE) if the document is requesting a payment or other call to action.

5 - Sealing Station aka Moistener, Closer, Sealer (MCSE)

This step requires the careful monitoring of the water reservoir to ensure that the inserter has the ability to seal all the mail pieces. Not sealing one or more envelopes in a run will affect the integrity of the job.

- sealing the envelope:
  - sealing can be skipped if the mail piece requires additional handling or if the job is for testing purposes,
  - this step would leave the envelope flap up if skipped, and

  - any exceptions to sealing would be programmed in the Insert Control File (ICF) or barcode, if only a few envelopes were not sealed they would be diverted for easy identification.

- envelope integrity:
  - although the document integrity is checked at various stages of the inserting process (either by the operator or more commonly by the inserter intelligence) the envelope integrity is also checked here using either:
    - cameras,
    - scanners, and
    - readers.
  - envelope integrity is typically checking the presence of a readable address in the window, the actual presence of an envelope, the ID number / sequence number of the document set and just like any other point in the process the system will alert the Operator or stop the inserter for further investigation if something is not correct.

6 - Weigh Station

The weigh station is used to apply the correct postage and / or calculate the exact weight of each mail piece for audit purposes. This unit;

- weighs each mail-piece,
- calculates piece count weight, and
- determines correct postage:
  - validates weight of piece against the projected weight in the insert control file (which was calculated by composition or the print stream manipulation program).

7 - Imprint Station / Postage Application

This station can apply the postage; if an indicia (postal service customer account number) is not used or in more advanced environments can imprint messages on closed envelopes using high speed black or full color inkjet heads, simplex or duplex. Imprinting at this stage can support for example using a standard envelope for all document sets.

- simple meter postage:
  - provides mailing date,
  - confirms mailer ID,
  - company indicia ('stamp' indicating customer account number to be charged), and
  - provides validation for comparison to the insert control file (ICF) for Electronic Statement of Mailing (eSOM)or manifest submission to the Post Office.
- envelope marking and / or envelope printing requires additional equipment and intelligence, to:
  - print on a closed envelope by using the ICF to drive the address or other file to match the envelope being presented, this could include:

        - mailer's:
            - □ logo,
            - □ return address.
        - recipient's:
            - □ address,
            - □ postal barcode.
    - custom (variable) messages to match the internal document set message, and
    - simplex or duplex printing can 'mark' the front and back of the envelope.

The inkjet systems today, designed for high speed variable printing at this inserter stage are designed to dynamically adjust the distance between the print heads and the mail piece meaning no additional set up is required to accommodate variable envelope thickness and can handle letter mail or flats (some overall thickness restrictions for flat mail will apply).

8 - Out-sort Station

The ICF can divert mail here for a variety of reasons, they include:

- 'pulls' / special mail pieces required for additional for manual handling,
- non-domestic mail, and
- postal-walk break marked with red line on the edge of a mail piece so that the Operator can easily identify the batch:
    - simply just stacks the completed mail pieces by split and the Operators move them to trays (or *lettertainers*) and pallets (or *monotainers*).

Speeds and Features

High speed production inserters, in the most common transaction document environments, range from 8,000 envelopes per hour to in excess of 25,000 per hour. High speed inserters from various manufacturers have a lot of features in common:

- they offer a low cost per piece (caveat being, with the right intelligence and processes in place),
- they handle a wide range of media / substrates,
- they are designed to work 24 / 7,
- they support their own or other providers workflow software for file based processing and document production management,
- they run close to rated speed when handling #10 and 6" x 9" envelopes,
- they run at slower speeds when handling larger envelopes (flats),
- an intelligent GUI for Operator interface,
- programmable job control,
- warning lights for alerting the Operator to an incident requiring intervention, and
- improved rotary and friction feed devices.

Low to mid-range inserters, not typically found on the production floor in the transaction document environment, may be found in a manual processing area. They range from 800 envelopes per hour to about 5-7,000 envelopes per hour.

Most will come without intelligence but will have a rudimentary counter / meter (to count the number of mail pieces) and features like double document detection. Most will fold both a C fold for an 8.5" x 11" document and a double fold for an 8.5" x 14" document and may only have the capacity for one or two inserts.

Some have the option to switch sealing on or off and some even have some level of intelligence added on aftermarket (such as an enhanced controller). It is important to weigh the benefits of a variety of models and understanding your requirements clearly before purchasing a low to mid-range folder / inserter because the differences can be subtle but the usability of the equipment can be greatly affected.

What is Document Integrity?

Document integrity is not only specific to inserting it involves all the steps in the document creation, build and production processes that result in accurate inserting and tracking processes.

Most commonly, document integrity refers to the bar codes or other mark/s on the page that support tracking, reconciliation and audit processes by confirming the individual pages are in the right mail piece and that the mail piece is complete and accurate and can be located or identified at any point in its lifecycle (OMR/bar code/unique identifying number that is either human readable (to facilitate manual collation and insertion) or machine readable to support mechanical collation and insertion.

Typically vendors /production management provide composition / programming guides, document style guides, stock and envelope manufacturing standards to support document integrity.

Document integrity must also be considered when planning for and executing re-makes (items damaged during production) and / or re-prints (items that may need to be re-printed by a call center or the recipient themselves.

It should be noted that production re-makes (sometimes called re-prints, although they shouldn't be) are one of the hardest items in which to ensure true document integrity, as often they do not follow the entire manufacturing process exactly as the main production run did. Quality of the image can be affected using a smaller / different printer, look and feel of the media / stock may be different (inkjet paper versus laser printer paper) they may come out of a post composition system wherein the composition system created the original integrity, and they most certainly won't be inserted by the production inserter as the typical re-make volume would not justify set up on the inserter.

Document integrity is not only about placing a readable (human or machine readable) mark on the paper it needs to consider the size and location of the mark / identifier on the document and takes into account the size of the document, the size of the margin or other white space on the document, the size of the window in the envelope, white space around the integrity mark to ensure readability, the types of folding in use to ensure there are minimal changes and impacts to document integrity should the envelope or document change, and also to support minimizing operator set up time and folding plate changes.

Integrity marks are not only found on the transaction document but can also be found on the insert and the envelope, this increases the overall mail piece integrity as it would ensure the right insert is being used with the document run and supports tracking of the actual mail piece.

Document integrity must also apply to electronic documents and electronic distribution as this is just as important as ensuring the right physical mail piece gets to the right person. Errors in electronic distribution are no less damaging and are usually recognized faster by the recipient, as access to the document is usually available before the physical mail arrives. All the same integrity considerations in document creation, document build, electronic production and tracking processes are required along with more sophisticated electronic security to ensure that the recipient is the only recipient who has access to their transaction documents on line.

Today's Key Players

Production inserting and/or peripheral equipment:

- Pitney Bowes
- Bell + Howell
- Gunther International
- Kern

Mid Production inserting and/or peripheral equipment

- Bell + Howell Pitney Bowes
- Neopost

Desktop (low to mid-range) inserting and/or peripheral equipment

- Pitney Bowes
- Secap
- Neopost
- Hasler Inc. (distributed by Neopost)

There are many other players associated with providing inserting and related equipment;

Many of the following produce / distribute low to high speed units (not often used in high volume transaction document arenas) or provide peripheral inserting equipment, such as feeders, card affixing, folding, cutting, packaging or reading/scanning technology that may support some transaction document production processes :

- GBR
- Lake Imaging
- Mailcrafters

- Mail-Quip
- PAC
- Profold
- Propack
- Sensible Technologies
- Streamfeeder
- UltraSystems

Basic Production Equipment versus Advanced Equipment

Basic equipment is as you would expect:

- standard configuration typically with one type of input station, either for roll input or for cut sheet input,
- may have between 4 and 6 insert stations, and
- typically has some job programming capability to support faster set up for known repeated jobs.

Advanced equipment, would include sophisticated systems such as:

- multiple input capabilities,
- many readers, scanners, cameras distributed at each station in the unit to constantly track the integrity of the document and the mail piece,
- job programming capability and ICF support for file based processing and full document management, and
- full color duplex imprint station integration as mentioned in the Description of Inserter Equipment Components and Sub-components section, 7 - Imprint station / Postage application.

Brief History of Inserters

There are, of course, more inserter manufacturers than are mentioned here. For the purpose of providing some technology history, here is a brief look at some of the best known (from a North American perspective) early pioneers of inserting technology, as this is not a history lesson about how many vendors there are in the world and who are best.

Early 1900's Bell + Howell were founded and made motion picture industry equipment, they were not in the mail sorting and inserting business until much later.

1902 - Chicago, Arthur Pitney introduced first double locking hand cranked mail postage meter machine and founded (via patent) the Pitney Postal Machine Company

1920 - Pitney merged with the operations of the industrialist Walter Bowes and began selling the first US Postal Department-approved postage meter. By optimizing postage meters, businesses could speed up the mail handling process and protect postage funds, while the post office could more easily track the money it collected from delivering mail. Today Pitney Bowes are reportedly the world's largest manufacturer of inserting equipment and dominate the tabletop inserting market (scaling down their tech for low volume use)

1923 - The Canadian Post Office gave its approval for the postage meter and Pitney Bowes established a Canadian subsidiary in Ottawa. Pitney Bowes Canada was formed.

1930 - Security was essential (and mandated by the government) to the success of the metered mail systems meaning that the Pitney Bowes record keeping had to be air tight and equipment could only be rented and never sold.

1936 - Albert Williams (founder of what would become Bell + Howell's inserting and mail machine business) invents, patents and produces the first automated mailing machine in Tatamy, PA (the first Phillipsburg mechanical inserter).

1945 - Directly after the end of World War II, Max Böhler and Ferdinand Weber founded the company Böhler und Weber – later abbreviated to BÖWE.

1947 - Marc Kern (founder of Kern AG Inc.) started the company in a small workshop in Konolfingen, Switzerland. Whilst Kern was employed in printing shops in Europe, he witnessed letter mail being manually inserted which led him to develop the idea of manufacturing a machine that could fold documents and insert them into envelopes, he built the prototype.

1953 - BÖWE SYSTEC produces its first cutter, starting the development of today's peripheral paper management hardware and solutions and in 1959 they introduced their first inserter system.

1959 - Bell + Howell purchased The Inserting and Mailing Machine Company and now they are in the 'inserting business'.

1960 - Bell + Howell opened a factory in Germany becoming the first international vendor or mail processing systems.

1961 - Pitney Bowes introduces their mail inserter.

1963 - General growth in business mail leads to the implementation of the 'zip code' in North America

1964 - Kern establishes an electronic department and the world's first fold-inserters (with optical reading capability) are delivered to their customers.

1960's - IBM's 1403 computer generates the first computer statements and bills and Pitney Bowes upgrades their inserter equipment to solid state electronics.

1968 - Pitney Bowes acquired the Monarch Marking System Company, which would produce the first barcode equipment for retail trade use.

1968 - Kern is on the leading edge of inserting innovation and introduces the first "online" system incorporating cutting, folding and inserting as a single process.

1969 - Hasler Inc. founded, specializing in low-mid volume folding, inserting, and mail processing equipment.

1971 - after Bell + Howell sold off some of their other business they made micro-imagery and mail sorting, the focus of Bell + Howell product development.

1971 - Bell + Howell develops the first intelligent inserters by adding reading technology (using Optical Mark Recognition (OMR) for selective inserts & greater integrity).

1971 - Postal OCR readers were introduced for computerized weighing and insert control programs.

1975 - BÖWE SYSTEC develops the world's fastest inserting system, with a speed of up to 12,500 insertions per hour.

1975 - Bell + Howell produces the first station-to-station matching system for the banking industry.

1976 - Pitney Bowes introduces a range of financing and payment options for their production equipment by founding their own Credit Corporation.

1977-1981 - Gunther International founded. Research shows several founding dates between 1977 and 1981. Gunther designs, manufactures, and services high speed production mailing systems to organizations in the United States and internationally and is best known for their innovations and technical leadership in high speed production mailing systems (mail inserting and ink jet). Also well known for its Champion software based around Microsoft technologies.

Mid 1970's - Secap was established and became a leading European Mailing Systems manufacturer and supplier leading supplier of mailstream management solutions, offering its extensive product line through a nationwide network of authorized independent and Secap operated local dealers, including postage meters, table top folder inserters and other mailroom equipment. Secap is a wholly owned subsidiary of Fimalac.

1982 - Bell + Howell produces the first inserters with computerized weighing, enabling greater postage discounts.

1984 - Kern launches the world's first modular inserting system.

1990's - Bell + Howell's Mail and Messaging Technologies, made up approximately 45 percent of the company's revenue, provided software-driven mail processing functions including collating, cutting, bursting, folding, inserting, scanning, encoding, and sorting. Its customers included organizations with high-volume mailing needs such as financial institutions, direct mail marketers, credit card companies, and others. The Information Access segment, comprised of Bell and Howell Information and Learning and Bell and Howell Publishing Services, was the second largest segment, bringing in approximately 36 percent of the company's total revenue (the split becomes more significant in 2001 and 2011).

1990s - Pitney Bowes continued to innovate in the field of digital technologies and software, introducing the AddressRight address & barcode printer (1991), the first in-line weighing and metering system (1992), the first secure digital postage meter (1995)

1990/1992 - Bell + Howell introduces a family of scanners capable of image enhancement technology, Adaptive Contract Enhancement (ACE).

1994 - Kern launches the first inserting system to achieve 26,000 mail-pieces per hour.

1995 - Introduction of the Automated Document Factory (ADF) concept.

1996 - Bell + Howell introduces the AccuFeed family of automatic document feeders for its line of scanners.

1996 - Pitney Bowes introduces the first ink jet postage meter.

1997 - BÖWE SYSTEC launches their high speed, high volume production level inserter.

1997 - Bell + Howell introduces their first ultrasonic multi-feed detection production scanners.

1997 - Bell + Howell acquires COPE Systems Inc., creators of high-speed inserters for handling large page count applications.

1997 - Kern establishes a direct presence in the US market with Kern International Inc. on their road to global inserting innovation and expansion.

1998 - Pitney Bowes introduces the Digital Document Deliver (D3) platform (1998) which provides message management via hard copy, web, email and fax.

1999 - BÖWE SYSTEC introduces their Turboplus inserter system, utilizing pinless technology and founded BÖWE Mailroom Communication GmbH, a subsidiary dedicated exclusively to software development.

2000 - BÖWE SYSTEC develops a new generation of high performance cutting technology to support high speed sorting and inserting.

2000 - Bell + Howell launches a new series of scanners with VirtualReScan (VRS).

2000 - The Kern 3000 Megamailer was officially included in the Guinness Book of Records as the fastest document inserting machine in the world.

2000 - 2007 - During this period, Pitney Bowes (considered the largest US outbound mail aggregator, providing mail pre-sort services and postal discounts to many organizations across the country), invested $2.5 billion in approximately 85 acquisitions (primarily in software and services businesses including MapInfo, Group 1 Software, PSI Group and Imagitas), expanding their offerings in mail services, intelligence software and printing. The Software Portfolio has been largely grouped into an operating division called Pitney Bowes Software.

2001 - Pitney Bowes acquires Bell + Howell's International Mail and Messaging Technologies, London, England; markets and services high-end mail processing, sorting and service-related products.

2001 - BÖWE SYSTEC introduces their high speed cut sheet feeder.

2001 - Bell + Howell establishes an enhanced line of inserters, advanced vision systems for complete document integrity, and their job tracking software (JETS).

2002 - BÖWE SYSTEC introduces their Turbo inserter, responding to the market's need for speed.

2003 - BÖWE SYSTEC and Bell + Howell combine North American operations, forming BÖWE Bell + Howell.

2003 - BÖWE Bell + Howell launches their color and bitonal/grayscale scanner series.

2004 - BÖWE SYSTEC presents its first high-speed inserting system, with an increased production volume of up to 14,000 insertions per hour.

2004 - BÖWE SYSTEC acquires remaining 60% share of its subsidiary, Roll Systems Inc., then enters an equal-share joint venture with LaserMax, a division of Stralfors AB , to form LaserMax Roll Systems AB, supplying high-speed paper handling systems for pre-and post-printing.

2004 - BÖWE Bell + Howell acquires Intellitech Innovations Inc., a longtime partner and provider of automated document factory (ADF) solutions, and its JETS and JETVision product lines.

2005 - BÖWE SYSTEC acquires an additional 10% of BÖWE Bell + Howell's holdings, now totaling 60%. The remaining shares purchased in further annual 10% portions.

2005 - BÖWE Bell + Howell acquires BCC Software, Inc., a premier developer of mailing software.

2006 - Kern's 3500 Dual-Channel Inserting and Mailing wins the iF (International Forum) award for the innovations and excellence in design.

2008 - Kern introduces the K2600 Inserting System, representing a milestone in the evolution of automatic mailing systems.

2010 -Gunther International and BÖWE Bell + Howell (BBH) announced an agreement whereby Gunther International will offer its customers disaster recovery and overflow services available through BBH, by marketing BBH disaster recovery services directly to its customer base.

2011 - Versa Capital Management (a private equity firm based in Philadelphia that invests in distressed companies) bought BÖWE Bell + Howell (The Information Access segment) and returned it to its original name.

2011 - Bell + Howell becomes ProQuest Company, comprised of the two companies formerly known as Bell & Howell Information and Learning and Bell & Howell Publishing Services. The international mail and messaging technologies business was sold to Pitney Bowes earlier in the decade and the North American high-volume mail equipment, software and services business remains as Bell + Howell.

2012 - Pitney Bowes is one of 87 existing firms that have been members of the S&P 500 since its creation in 1957 and they are well known as an industry advocate, they have focused in recent years on advancing postal reform, the USPS Intelligent Mail barcode, 'TransPromo' and environmental sustainability.

References

Some images are from acadami™ course materials courtesy of relevant vendors.

Some of the history is courtesy of acadami™ course materials.

Some of the history is from web based research in the public domain.

Fold images are courtesy of Trish Witkowski at FoldFactory.com and are used with permission.

Document Delivery Process

Contributed by Brett Dashwood edp, William Broddy m-edp

What is the Document Delivery Process?

The postal service is a unique entity. It is an independent agency within every country that is charged with the secure delivery of private and personalized information to every known postal address within the country. Every nation has a postal service, many of which are bound by the rules of the Universal Postal Union[1]. The UPU's role is to allow the speedy delivery of mail both inside member post offices and between them. It also develops standards for postal delivery such the Four State Barcode,

Postal Law

Both statutory and common law give the post a special status under law and expect of it certain responsibilities.

Mail Box Rule

Whether it is called the *mail box rule, post box rule, post office rule,* or simply the *postal rule,* there is a presumptive rule in most countries that is a part of the legal rules of evidence. Courts presume that a document[2] placed into the postal system will be presented to the intended recipient[3]. The issuer of the document only has to provide proof of a chain of custody from the beginning of the creation process through to the posting of the document at the post office.

More recent citations require proof that the posted document has the correct address and adequate postage. If these criteria are met, it is presumed that the document has been delivered and the onus is on the recipient to prove that it wasn't.

The second part of the reule is that mail with the correct address and postage cannot be returned to the sender by the post office, even if requested. Once mail has been posted, it is gone.

Right of Way

"Neither snow nor rain nor heat nor gloom of night stays these couriers from the swift completion of their appointed rounds."[4]

The post office has the right of way to any legal postal address. This means that no one can stop a postal employee or vehicle from delivering mail. The right of way rule also

1 The UPU is a special agency of the United Nations with 193 member countries plus Vatican City.
2 Most postal agencies restrict this to First Class mail.
3 Example of the Mail Box Rule being cited in court rulings: No. 10-5205. United States Court of Appeals, Sixth Circuit - SOLOMON OLIVER, Jr., Chief District Judge Filed: October 6, 2011: For the presumption of receipt to arise under the common law mailbox rule, a party must present evidence that the letter was "properly addressed, had sufficient postage, and was deposited in the mail".
WIGMORE ON EVIDENCE states: "The presumption [of receipt] rests upon the supposed uniform efficiency of the postal service in delivering letters duly stamped, addressed, and mailed into its custody; if therefore the efficiency is operating, does not the non-arrival of an alleged letter indicate that such a letter was never given into the postal custody?"
4 Although the Creed of the US Postal Service, it actually is a translation from Herodotus' Histories, describing the Persian Imperial courier service circa 500 BC.

means that no other government agency[5] (or other 3rd party) can interfere with, hold, or divert mailed documents, except under court order. The post office is obligated to deliver it only to the addressee or their authorized agent. Also implied is that no one, other than the addressee or their agent, has the right to open the mail that has been delivered.

Both the *Mail Box Rule* and the *Right of Way* doctrine ensure mailers a secure, efficient, legally accepted way to present documents to the recipient that is significantly less expensive than any alternative form of physical delivery.

Mail Fraud

Most countries classify mail fraud under more general fraud laws; however, in the USA mail fraud is an explicit crime.[6] Most legal jurisdictions will more readily recognize intent if a crime is committed via a posted document. Most postal authorities have the right to refuse mail that they suspect of being misleading.

History of Mail and Postal Organizations

Postal agencies have existed for millennia, with examples of mail pieces dating back 2,500 years. Most, however, were couriers of government documents. *Cursus Publicus*, the Roman postal system, was the first independent postal service that carried documents for both government and private citizens. Its rules form the basis for the both the *Mail Box* and *Right of Way* rules.

Prior to the modern Post Office, with postage paid by the originator and universal delivery, postage was paid by the recipient. The world's first pre-paid stamped letter sheets were introduced in Australia in 1838, followed by the introduction of adhesive postage stamps in the UK in 1840.

With the migration to postage being required up-front, the amount of uncollected, unpaid for mail was removed from the delivery system, dramatically lowering the cost of delivery. For example, in the 1840s the postage cost of a document sent from New York City to Buffalo dropped from 25¢ to 1¢.[7]

Business Mail Preparation Requirements & Costs

Although the specifics vary, most countries have requirements that a mail piece must meet to ensure that it gets posted (or mailed or lodged) as business mail to ensure that the best postage discounts are available. If the specific requirements are not followed, it impacts the postal rate for the mailing.

The discount structures in place for most countries are to make things easier or cheaper for the post office. Understanding a particular country's requirements for postal discounting is very important. Typically these requirements include sorting and barcoding, as well as some form of grouping or batching. These requirements are designed to cut out certain processes or to minimize human interaction with the mail where possible.

5 Not even the National Security Agency (NSA) in the USA.

6 In the USA, fraud is a state criminal offense unless it crosses a state border or is committed via the mail. Mail Fraud is a federal offense.

7 It is amazing how profitable a postal system can be when it clears out, or properly charges for, all items that get delivered virtually for a lot less than first class mail, such as unaddressed advertising literature and magazines.

Postal sorting is typically based on the postal code or the ZIP code as appropriate, which designates a smaller geographic area, such as a suburb or town. This helps the post office by grouping the mail so that it can be more easily distributed in bulk, closer to the point of final delivery.

Regardless of whether you mail a letter to someone in the next suburb, another state, or another country, it must go to a centralized distribution center to be sorted. Each item is then forwarded to another distribution center, or to the appropriate post office or sorting center.

In a Big Country

In geographically large countries, the logistical costs involved in delivering a piece of mail next door or to the other side of the country are very different. Some may involve only road transport; however, some may require air travel. Some countries deliver by helicopter, dog sled and boat. Most countries amortize the logistical cost of delivery and have a single cost within the country regardless of distance. In other countries there is a user pays system, where the postage rate changes based on distance.

Some countries, like Canada, offer virtually no discount. However in Brazil, because approximately 85% of the business mail for over 200 million people is generated from in and around 3 cities in the southeast part of the country, they have different rates for within state (local), bordering state (regional), or elsewhere (national). There are 26 states and 1 federal district in Brazil.

Another example is Australia. Although the difference in price is not as proportionally different as Brazil, business mail sent in bulk within the same state is about 3% cheaper when compared to mailing to other states. However, that is only if it is sorted into groups of 300 items going to the same distribution center. In Australia this is called a direct tray. For all items going to a distribution center that do not fit into a direct tray of 300, regardless of destination (same or other state), even if correctly sorted and barcoded, the postage rate may be 7 to 17% more expensive.

Size, Shape, and Weight

Trays, bags, and boxes of mail, are designed to carry a certain quantity or certain weight of mail. Looking at an individual mail piece you may not think it matters that much whether it is 50g (~1.7 oz) or 200g (~7 oz), however when we look at 1,000 similar pieces, the difference of 150g is now 150kg (~330 lb). If we now look at 100,000 pieces, then the difference in weight is now 15 tons (~33,000 lb). Size, shape, and weight requirements can drastically alter the postage amount required to deliver a piece of mail.

Here is a table of information from Australia Post about the different sizes and weights appropriate to different mail classes.

	Small	Small Plus	Medium	Large
Max. Weight	125g	125g	250g	500g
Min. Size	88 x 138mm	88 x 138mm		
Max. Size	130 x 240mm	162 x 240mm	180 x 260mm	260 x 360mm
Max. Thickness	5mm	5mm	20mm	20mm

	Small	Small Plus	Medium	Large
Shape	Oblong	Oblong	Rectangular	Rectangular
Common Examples	DL (110 x 220mm) C6 (114 x 162mm)	C5 (162 x 229mm)	B5 (176 x 250mm)	B6/C4 (125 x 324mm) C4 (229 x 324mm) B4 (250 x 353mm)

Square articles are permitted in the Medium and Large size categories, but are not permitted in the Small and Small Plus size categories.

In some countries the weight of the smallest mail piece, with the smallest postage rate, may not exceed 50g (~1.7 oz.), with some as low as 30g (~1 oz.).

Article Type

Most business mail is an envelope containing a document, either:

- plain faced with addressing details directly applied, or
- window-faced where the addressing details are visible through a clear plastic window in the front of the envelope.

The document itself could also be the address piece, in one of these formats:

- **A self-mailer:** Where the document is folded and glued to enclose the information being mailed.
- **A postcard:** Where the document is a single piece of printed card of a size and shape appropriate to the message.
- **Plastic wrapped:** Plastic wrapping of articles can provide an alternative to envelopes. Artwork and delivery address details can be printed on the document itself, or a sheet of paper called a flysheet which is visible through the plastic wrapping. Plastic wrapping may have adverse effects on the ability to lodge mail at a specific postal discounted rate.

Base Stock for Envelopes

The paper or plastic base stock used to create envelopes is important to keep in mind and typically must meet minimum country-specific criteria.

Each country has different requirements for the color of envelopes, postcards, or other articles. The reason for color requirements for envelopes is usually because of the machinery that is used to process the mail. Choosing light or pastel colored envelopes are usually preferable than dark, solid colors, but checking local postal rules is always advised..

Article Flexibility

Business mail must be reasonably flexible to ensure that they can be processed through high-speed letter sorting equipment.

Article Layout

Each country has similar, yet different, requirements around placement of certain pieces of information. In addition, there are areas that must be clear of information called quiet zones.

Addressing of business mail usually consists of different zones with different requirements; typically they are referred to as:

- Return Address Zone
- Postage Zone
- Delivery Address Zone
- Postal Barcode

Return Address Zone

The position and dimensions of the Return Address Zone are not as restrictive as those of other zones. In most countries the return address is located either in the top left on the front or at the top on the back of the mail piece.

When printed on the front, the dimensions and size of the zone are taken from the top left hand corner of the mail piece. When printed on the back, typically the return address must reside within a certain area from the top of the mail piece.

The return address must be legible to the naked eye, but must also be small enough not to be mistaken for the actual address by a person or machine. Typically the return address is printed in a font no larger than 8pt and usually not in a bold font.

Postage Zone

For business mail, the postage zone contains postage indicia of one of the following types:

- Postage Paid imprint,
- postage meter impression, and
- Ccstom or personalized imprint or indicia.

The position and dimensions of the indicia are restricted. In most cases globally, postage indicia are located in the top right of the front of the mail piece, and the dimensions and size of the zone are taken from the top right hand corner.

Delivery Address Zone

The Address Zone is typically the largest zone. Usually a defined maximum size is given (based on reading equipment) and its position is usually determined by quiet zones on both sides of the zone and also to the bottom edge of the envelope. These quiet zones are sometimes used by mail processing machines to mark ends off batches or a sort group, etc.

The Delivery Address Zone can be larger for non-standard sized business mail, as the pieces typically do not go through the same high speed mail processing machines.

The skew of the position of the address zone varies between countries; it is typically ±5 degrees to – 15 degrees from parallel.

Each country's postal authority has similar but different requirements for the layout of the delivery address within the zone. For example, the requirements for domestic (within country) mail in Australia are:

- left justified,
- bottom line contains the locality (suburb), the state or territory, and the postal code,
- second to last line contains the number and name of the street or thoroughfare or post office box or bag number,
- other address lines, such as a person and/or business name, are classified as simply additional address information which are placed above the last two lines of the address and not below, alongside or within these lines, and
- the postal barcode is then placed above the address block within the delivery address zone.

For business mail, the delivery address zone should be positioned parallel to the long edge of the mail piece. For non-standard mail, the delivery address zone should be positioned parallel to an edge of the mail piece, either portrait or landscape.

Different countries also have different font and font size requirements.

Using Australia as an example again, any font can be used for the delivery address as long as it is clearly readable, preferably in 12 point size. It is also strongly recommended that print characters do not overlap. While attributes such as italics, bolding, shadowing, or underlining are permitted, their use should be avoided. Dark colors such as black, dark blue and dark green are preferred. Red, orange and yellow inks should not be used.

Other countries may specify a typeface or family of fonts that are required, and in most countries there are markings that are not permitted on the front of a mail piece, such as fluorescent ink or designs that may be mistaken for postage stamps, franking impressions, other indicia, or service labels. Terms that look like a post office service, such as priority mail or express mail are not permitted.

Logos, advertising content and other printing may be added on the front and back of mail pieces, however when allowed, they can only be used if it does not obscure the inscriptions in the postage zone, return address zone, and the delivery address zone.

Although it is technically part of the Delivery Address Zone, the postal barcode typically has its own print, size, positioning, and quiet zone requirements.

Postal Barcode

A postal barcode is a machine readable representation of information, usually printed as parallel lines, that improves the speed and accuracy of processing mail. It helps in sorting the mail faster as its content is typically encoded delivery, routing, and/or tracking information. The postal barcode can also be used to provide information to mailers, either prior to lodgment or for the management of returned mail.

Four State Barcode (4SB) is the *de facto* standard for postal tracking. This barcode consists of a series of vertical bars, each representing one of four possible states.

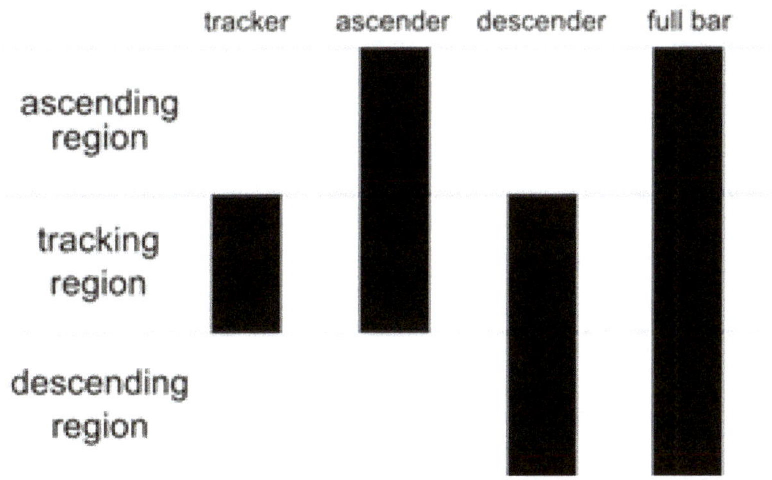

The Four State Barcode is used as a postal code by:

- Australia Post
- USA for the USPS IMB
- Canada Post
- Royal Mail in the UK
- Japan
- Singapore

Interleaved 2 of 5 is another barcode format. Often called "2 of 5" it is used as a postal barcode by Deutsche Post in Germany and Italian Post.

The Universal Postal Union (UPU) has a number of standards covering mail and addressing requirements; two that are of interest:

S18e Part E provides a specification of the encoding of UPU ID-tags using 4SB symbology that may be applied to the front side of small letters.

S19d Part D deals with the definition of areas that are, or can be, used for the printing of addresses and customer applied barcodes which are printed together with the delivery address.

Additional information on these and others can be found in the UPU standards catalogue on their website at www.upu.int.

Change of Address Register

Many countries have a National Change of Address (NCOA) register. This is usually available to mailers to help ensure maximum accuracy for delivery of mail. Any discrepancies are typically reported back to the document generator/mail owner rather than altering the actual mail piece.

Lodging Mail

Prior to submission or lodgment to the post office, some form of report must be completed, such as a statement of mailing or manifest. In countries where it is appropriate, most postal sorting products generate these already populated forms and today they are also usually available for electronic submission.

Details on a manifest report include quantity, special handling items, international mail, and items with special distribution requirements.

Quantity

To qualify for different postage rates and discounts, most countries require mail to be submitted or lodged in a single delivery grouping of a certain quantity. That quantity may be as low as 300 (eg. Australia) or up to 5000 or more, depending on the country and the postage rate desired.

There is also a common requirement that s specific percentage of the delivery grouping must be readable by the mail processing equipment and/or have a correctly formed and readable postal barcode.

Special Handling

Not all mail in a single production run is able to go through as standard business. Special handling may be required for physically larger mail pieces or pieces that require different packaging. This packaging may be a gusseted envelope, a courier envelope, shrink wrap, or maybe even a box or carton. These special items usually require manual handling (and sometimes addressing). They often require a separate submission to the post office or a different delivery method (eg. courier).

International Mail

International Mail must be included separately on the lodgment or manifest report. These items may need to be packaged/bundled separately than those for national delivery and may need to be handled by an international mail consolidator.

Special Handling

Most countries have a special handling category that covers items with special distribution requirements. These requirements may cover:

- bundling items to be sent to a branch or a broker,
- different sorts,
- special bundles may require rubber bands or special envelopes,
- items too big for machine inserting and that must be manually handled, and
- special inserts or no inserts at all.

Because these items are typically handled manually outside the normal production process, privacy needs should be considered when looking at the environment in which they are being processed. These items may not get mailed individually with the post office as they may need to be delivered to an address that is different from the actual delivery address on any of the mail pieces. For example, property levy notices being delivered to a single property management company rather than to the individual property owners. This may mean a different sorting requirement and a different split of production files.

If the items are significantly larger than normal post they may go with a courier or through the standard post office parcel service. In this case a separate system needs to be in place to identify the distribution method and address, as well as the creation of postage/ shipping labels or courier slips, and any separate tracking requirements.

Matching Documents from Different Sources before Mailing

A special area of Special Handling is when different personalized items need to be produced in separate processes and matched to be mailed together.

An example is credit or other cards and their carrier letters (or simply carriers). There are a variety of plastic card matching systems around the world from a variety of vendors. To effectively and accurately automate this process it can't simply rely on items being produced in the same sort order. While this helps, machinery-based matching must occur, too. The most common way is by a barcode printed on the carrier, either a linear or a matrix barcode, and having the cards in a queue or a physical buffer that are read and matched to the carrier to which they are glued or tipped.

A similar process is often used for matching documents that have come from two separate systems, such as a personalized document run being matched with copies or scans of completed application forms.

Small Volumes

In the world of business mail production, one of the most overlooked areas is small quantity jobs. Although the main statement run or runs may make up 90% of the volume, the last 10% of the volume may be from 90% of the individual jobs to be run.

These smaller volume jobs are often overlooked in the requirements gathering stages of big projects. Some simply include their quantities in the calculations of the main runs, however the logistics involved may mean that these documents could cost 10 times or more the price of a standard statement for example.

When gathering requirements for large projects, make sure that all document types and their processes are completely understood and documented. Make sure that the right people are asked the right questions, which may mean asking in a different department than the main document type (eg. operations instead of marketing).

One area of importance is the impact that these large numbers of small quantity jobs may have on Service Level Agreements (SLAs). Once understood and documented, they should be designed and an appropriate test platform built to ensure that all processes and times are known.

This book is set in Century Schoolbook 11pt.

Headings are in Sans Source Pro.

To access some characters Arial Unicode MS was added in some text blocks.

Book layout by McGrewGroup /MC2 Books, Aurora, Colorado

Printing at Hudson Printing Company, Salt Lake City, Utah

www.ingramcontent.com/pod-product-compliance
Ingram Content Group UK Ltd.
Pitfield, Milton Keynes, MK11 3LW, UK
UKHW051424130926
15127UKWH00014B/103

9 781893 347052